Other books

Empty Thoughts from an Empty Head
Spaced Out and Cut Up
Stupid Jokes for Clever People & Clever Jokes for Stupid
People
That's the State We're In

~~~~

Books in the Logic List English Series

Logic List English – Rhyming Words etc. - Volume 1A
Logic List English – Spelling Arrangement – Volume 1B

Logic lists English - Multi-Syllable Words - Volume 2A
Logic lists English - Split Multi-Syllables - Volume 2B

More Volumes Coming

Observations From
Another Planet
(Aphorisms & Axioms for the Modern World)

By
Tony Sandy

DragonEye Publishing

Observations From Another Planet
(Aphorisms & Axioms for the Modern World)
Copyrighted © 2017 by Tony Sandy

First Edition
First Printing May 5, 2017

ISBN 13: 978-1-61500-148-4 (Paperback)
ISBN 13: 978-1-61500-127-9 (ePub Ebook)
ISBN 13: 978-1-61500-181-1 (PDF)

Library of Congress Control Number: 2017940165

Published by
DragonEye Publishing
511 W. Water St., Unit E
Elmira, New York 14905 USA

Website: DragonEyePublishers.com
Contact: Orders@DragonEyePublishers.com

Contents

1   Preface
2   The Systemic Universe
98   Part Two (The Politics of Experience)
125 Life-Laws
126 Afterword

PREFACE

My life is my lab and my
proof is to found in your lives

Man, Know Thyself
(The AncientGreeks)

The Unexamined Life
Is Worth Nothing
(Socrates)

THE SYSTEMIC UNIVERSE

I have a theory that basically everything works on what I call an Accumulation/ Discharge Cycle. In karmic terms this is known as The Pendulum Effect and in purely physical terms as the vacuole effect. This includes the tides, seasons, emotions, sex, life and death as a single unit (positive to negative poles), thought and action (again as a single unit), stress and release (relaxation), volcanic activity, earthquakes,pulsars, vibration, the body's circulatory and digestive systems, black holes and white (w)holes, time - that is movement of attention (energy) from past through to the future, across the medium of the present, space as movement of matter across a gap (here to there) etc.

I call this also The Systemic Universe or Two Worlds Exchange (The Tao (Way) in modern terms). This creates equations of equal but opposite realities or opinions in human terms. Thought stops action - action stops thought or when viewed as a continuum of human endeavour / perception of existence as a chain, it becomes a scale or grade e.g. past, present and future/ colours - white, yellow, orange, red, brown, green, blue, black. War and peace even follows this path.

What we understand we can control. What we cannot control leaves us feeling frustrated and helpless (a victim of a greater reality - bigger than we are as individuals): Think of learning anything, be it in a classroom or workplace. When things become too complex for us (mysterious/ inexplicable), we panic and attack them, thereby reducing them as a threat to our ego, by simplifying our existence. This state of insanity is simply the inability to understand and therefore cope with changed circumstances (trauma - the shock of the new/ different). This is the seat of all prejudice: The only thing we ever truly fight is our own ignorance.

Why does the unknown (paranormal/ new/ different) frighten us? Because it is the tap on the shoulder that tells us we are not alone in the universe (The only form of life or being there is) and that we don't know everything there is to know about reality, which reduces our status to that of a helpless child again, instead of the pretentious adult we all like to think of ourselves as (The Emperor's New Clothes Syndrome): This is why death scares so many people too. It wakes us from the dream of the ego that tells us we are God and reminds us that we have mortal remains as well (All pilgrims on the road to The Undiscovered Country - "Come in number six, your time is up!" It is the bigger self reminding the smaller self, that there is more to life than this little pinprick in time and space, we call present reality (Star Trek, "The Squire of Gothos", TV episode).

Courage (learning/ enquiry) goes with not knowing. This is because if you don't 'think' you already know what's going on, you'll investigate to find out the truth. However if you do think you know the answer already, you won't look because you're sure you have the truth already (Prejudice is the full stop on your life as innocence is the door).

Intelligent people face their fears (study them - the scientific approach). The frightened try to eradicate what they're scared of (Suppress it/ kill it off - out of sight, out of mind - the political alternative). The first learns what makes

3

others tick as well as themselves - the second doesn't want to know how anyone thinks, including themselves as it would mean being response-able for their own existence and they'd rather somebody else was (Their parents, partners, society, God).

Why is a leap of faith so frightening? because it means throwing caution to the wind and abandoning everything you've ever known, owned or believed in (trusted): Every thought, every possession. Nobody can 'try' for us and in the final analysis, the experts (Those who've gone into a field before us) can only give us advice and information but the final gamble (decision/ choice) has to be ours. The future (heaven) beckons - are you ready and willing to jump into the unknown?

The reason that we suffer terror/ joy when the unknown enters our lives is down to joining a different world/ reality as a spy (plant) originally and having blended in so well, that we think of ourselves as genuine members of that society or group, until disturbed by the other side again (ghosts, UFO abduction phenomena, the witness protection scheme - all suffer from this denial of an old life, pre-birth or walk-in agreement, or the existence of an alternative reality [spirit world/ other dimension]). It is what I call the Sleeper Effect but is also allied to The Stockholm Effect because it is about losing your old identity and adapting / adopting to a new one, even if unconsciously (see below).

We think things are insane when we don't understand them and think they're sane when we do i.e. see the reasoning behind them (It is this that converts us from anti to pro something and vice-versa - in other words, simple belief changes how we see everything).

Our beliefs are dangerous - not because they are 'necessarily' right but because we act on them as 'though' they were (and that we have the 'right', morally, to do what we do, based upon this belief in our being correct in our judgement and attitude towards the matter in hand). To act or not to act, that is the question. The more intelligent think things through in depth (Question their perception / the evidence, plus their own motives i.e. the results and why they want them) before acting, whereas the less thoughtful, react before getting to the bottom of things (finding the truth) or questioning their own actions.

Seeing limit, we despair at finding answers/ surviving - this is why we need 'God' (eternity/ infinity) in our lives. Death of the spirit is giving up hope. This is displayed in terminal depression - that is giving up the effort to build or control, socially or individually, thereby destabilizing our lives as individuals/ society as a whole.

We become violent deliberately, to destroy all links with that which we've run out of patience with, so that we're not tempted to engage with it again (Eradicate it and we don't have to communicate with it i.e. learn of its existence again or anything about it and vice-versa with regards to its learning about us (mutually exclusive agreement to deny the existence of each other, to ourselves and third parties - The Ostrich Effect or Emperors New Clothes Syndrome): Free exchange of information / materials comes with understanding - resistance comes when this isn't mutual.

Observations from another Planet

When we're in ascendency, we're on the path of material gain (affirmation of life/ peace/ creativity) - that is letting go of old beliefs. When we're on the descending slope, we're in denial because we don't want to admit to ourselves (or anyone else) that we're losing our hold on life (loss of control/ feelings of weakness, displayed as acts of violence). Gentleness is therefore strength because it displays generosity of spirit (see 'God' quote above) as violence is just an act of despair (distrust/ abandonment) - be it by a country or an individual: Violence is frustration at our own ignorance and therefore helplessness because what we don't know (understand), we can't control. What do we fear most? Change because it betrays us all as blundering incompetents, which is a blow to our pride (arrogance is thinking we know as humility is discovering we don't). On the sensory level this is displayed as 'vision' (tension/ elation/ proximity (detail)) and 'sound' (relaxation i.e. letting go and therefore drifting drifting away from something you've given up on, that is depression as a state/ generalization).

Patience is freed attention (forward motion available through time or space) - impatience is attention held in thrall (drawn within to thoughts (questions) about the past). This movement' of attention, within and without is known as The Artist Effect or stepping back to observe and forward to act (intellect to see and emotion to feel i.e. enjoy your endeavours, if that is the 'right term'. There is a fear/ intelligence connection in all this or thought/ action ratio. The urge to get away/ get done with things versus a willingness to stay and explore (experience) what is here.

Observations from another Planet

Rapid movement leads to short cuts and simplicity but can also lead to half done actions. Patience leads to development into more and more complex forms because time is applied to the equation, rather than 'no time' (panic). In short the long way round is thorough and builds gradually, step by step, as impatience destroys because it is always rushing to break connections and get away. This also explains simplicity of speech (short hand/ slang/ use of single syllables) as opposed to complex, detailed explanations (long hand/ technical speech/ multi-syllable words). Lastly, rapidity of movement (bodily processes/ metabolism) wakes us up as depth of thought puts you to sleep.

The intelligent are active because they are curious about everything (investigative minds, driving their bodies eternally forward). The unintelligent tend to be docile and accepting of things as they are, not experimental rebels into finding out all they can about everything. This is why the latter are usually good natured and the former can be given to violent moods from frustration (Don't like obstacles/ fight to overcome them). If you relax, you can expand and let things fall back into place (Go back to how they were/ cure yourself of old wounds i.e. scar tissue/ ridged belief systems).

Movement is the teleportation of matter across space, through expanding it into energy(warmth i.e. e-motion or physical movement), then cooling it back down into matter, followed by settling (Forming i.e. inward movement/ thought): Laurie Anderson song ''Walking and falling' (transmission and reception).

Observations from another Planet

What exhausts us is manoeuvres. It is the start/ stop movement, especially when it is small and detailed, that takes it out of us. Only settling down and in, fills us up with energy as movement up and out, disperses it, draining us.

According to science, politics and education there is only one answer, one correct way of doing things. According to the spiritual path there are many routes to the same destination - artistic, healing ways (abandonment of the past/ separation from fanatical certainty into sceptical disbelief in anything (freedom from).

God is 'I will'(can) - the Devil is 'I won't' (can't).

Patience is freed attention (forward motion available, through time or space). Impatience is attention held in thrall (drawn within to thoughts (questions) about the past).

The simple complicate things to appear intelligent. The truly intelligent simplify things to 'be' clever.

God is any superior power we surrender to and learn from.

We become The Devil (evil) when we resist opening up and learning from those who know better from experience ,'what is here and how it works' than we do, because they created it or because they are older custodians of it than we are.

We make God (good) in our unity and The Devil (evil) by our disunity.

8

Observations from another Planet

Nobody likes to be humbled by their own ignorance (return to childhood state) but it's the only way to open up and learn about something new and different (anger, rebellion and frustration (resistance) are fighting the process of assimilation (change/ adaption).

We're all masters of our own universe (Gods). It is boredom that leads us into exploring beyond our own limits, where we become mere mortals in blunderland (No longer all powerful but all weak, all ignorant, all helpless).

Gratitude allows us to learn and move on - resentment keeps us trapped where we are ('Eternal outsiders, who don't know the entrance password). 'Better to serve in heaven than reign in hell ' to misquote Milton.

To learn (affirm) is to open up and gain confidence (feel significant). To ignore (deny) is to shut down and feel inferior (insignificant).

The reason we're reluctant to commit ourselves is that we're afraid of appearing less than perfect (not knowledgeable or adept at something). So by doing nothing, saying nothing the illusion of infallibility is maintained through silence, not tested by experience (chance). It makes us feel powerful to help others and powerless when we can't help them or even ourselves.

Time is spiritual wealth.

God allows - The devil forces. What this means is that good encourages you to act by giving you time and space to think and make your own decisions, whereas evil tries to panic (push) you into reacting i.e. leads you into ill thought out actions, that you may not survive or that you regret afterwards. The alternative to this is putting the brakes on because you suspect the motives of the person/ people trying to force you into action.

We can forgive what we can understand. What we have no understanding of, we have no sympathy or time for (our ignorance makes us angry about it i.e. defensive)

We always become obsessed by that which beats us and lose all interest in that which we can beat easily. This is what turns into mean (in both senses of the word) and nasty addicts, in pursuit of something or the disaffected, which abandon the same thing, when it no longer holds our interest (Is no challenge anymore - an inferior existence as opposed to superior one : War is a battle for our attention by what is in awe of us as peace is maintained by those indifferent to our lives (death/ no game situation).

Understanding is that 'Eureka!' moment that frees our attention from the mystery of how to interact with something as anger is the realization that we cannot control something (Shut out by our own ignorance of the rules the game - that is no ability to operate a vehicle of communication that exists on some level/ failure to be authentically alive (whole/ perfect) in some area of reality).

Outsiders are violently trying to break in (join) as insiders are peacefully leaving (abandoning). This is the explanation of conflict within us and its manifestation without (cause and effect/ growth and distortion/ birth and death). Contact with something we don't understand (hate/ fear) is like salt on a slugs tail - it causes us to shrink back from it. Because it makes us feel small or non-existent, we try to destroy it or shrink it down to a smaller, less intimidating size. What we're not afraid of (understand), we allow to grow/ expand.

What we're expert in we tend to be frivolous about but what we know nothing of, we take seriously.

Love is a measurement of what you lack in life - hate, what you have too much of and boredom, what you've had enough of (empty/ full/ overflowing).

What we concentrate upon, we're aware of - what we disperse (move our attention) from, we lose perception of (ignore the existence of): It's not that it doesn't exist, just that for us it seems imperceptible because of our time/ space relationship with it (It isn't there because we aren't there with it as it would be visible to us if we occupied the same continuum as it: Just because your body is here, doesn't mean that your mind is focused here too).

When we're tired, our perception of our abilities and indeed the abilities themselves, slow down so that we achieve less. This is because of entropy. Without rest we cannot gather enough energy to kick start our lives , with the force to sense/ achieve a lot within a little time/ area.

Contrast sharpens our perceptions and allows us to make comparisons, where continuity of effect (white-out conditions/ sameness) blunts them, so that separation of awareness, of distinct entities, dissolves into blindness of form (homogenous blurring of one thing into another, be it internal thought or external reality).

What we know, we can control - what we don't know, controls us (walks all over us). We build upon our successes and are destroyed by our failures (infinite growth versus eternal collapse - crawling out of a black hole as opposed to falling back into it).

We abandon what we cannot understand because we can't stand the pain of failure (The humiliation of not knowing and therefore not being able to control something). It is easier to destroy all communications with something than build or maintain them (trust rather than distrust)

What you're an expert in, you're more than happy to share your knowledge or expertise with. Our ignorance and incompetence cause us to want to withdraw from contact with others and hide because we're ashamed to display our inabilities, instead of exploit them as clowns (pride thing).

We destroy things around us to simplify our lives, when we can't deal with new input and build upon them in those areas where we shine (Flower in our strengths/ complicate our lives with detail).

A positive attitude leads us into regret, through the results of our actions upon others as a negative one creates it.

We run off into the outside world to avoid facing our thoughts inside (conflicts) i.e. the learning zone. Rather than deal with our ignorance, we try to find solace for our internal failures through external distractions (feel sorry for ourselves instead). Only the acceptance that nothing really matters (failure/ death) saves us from this addictive retreat , away from life (learning/ challenges to our ego i.e. pride in what did know and could do, not what we can't do and don't know now).

It is attitude which decides our reaction to life situations and how we handle them i.e. intolerance and rebellion against them or acceptance and learning from the situation (love or hate).

To hate is to reject because you don't understand and therefore feel a victim of. To love is to enjoy (accept) because you want to understand (learn from) i.e. don't feel a victim but want to become a victor over your own ignorance of something.

When our lack of knowledge is betrayed by our inability to interact with something, we withdraw contact from the world (Understanding frees our attention, so that we can once more enter the world as an active participant).

Patience is giving yourself the time to do something - impatience is tearing yourself away from contact because you don't believe you have the time to spend on this project.

We override the rights of others, to do things because we believe that we are more right than they are in an argument or have more right to act, morally or historically (ownership/ justice).

When we think we're going to live forever, putting up with an unchanging world, we want to die. When we think we 're going to die and lose the chance to explore the unknown world around us, we want to live.

When we feel guilty about our attitude to someone, we tend to ignore their existence. If we have no hard feelings towards someone or something, we're more likely to be open and affable towards them.

Confidence is an automatic response to a situation. Lack of confidence manifests in feelings of inadequacy ('I don't know what to do!'/ 'how to handle this situation') or worries over time ('Not now!'/ 'I can't fit this in'). Confidence isn't distracted - it lets things wash over it, without reacting (Lack of confidence is fear manifesting through resistance).

Our beliefs are dangerous, not because they are 'necessarily' right but because we act on them as 'though they were' (and that we have the 'right' to do what we subsequently do, based upon the former belief i.e. right and righteous).

Entertainment is getting away from it all (forgetting your troubles/ existence). Information is getting stuck into it (working your problems/ living your life/ putting in effort).

To act or not to act, that's the question? The intelligent think things through in depth (question the evidence/ their own motives) before acting, whereas the less thoughtful react before getting to the bottom of things (discovering the truth) and without questioning their own actions.

Observations from another Planet

Hope is seeing some new avenue of approach - despair is seeing none (eternity and infinity keeps us going - here and now stops us dead in our tracks).

Violence is the simple answer for materialists because they always want the easy way out, not the difficult way in (thought/ consideration). They attack things because their way of life is an assault upon it anyway, aimed at wearing down opponents/ opposition but which also wears them down too (input negated by severe output).

All discoveries are accidental because deliberate acts are defenses, aimed at shutting things out as opposed to letting awareness in.

All accidents are caused by being too close or too far away to contact and interact with things adequately (overdoing or under-doing it).

When we're focused on a goal, nothing and nobody else matters or exists, except as interruptions (unwanted) or diversions (wanted breaks). Socialization is the opposite of individuation - your attention is dispersed and you resent efforts aimed at getting you to respond (act/ concentrate upon something new). When you're involved in mind games, the body doesn't exist and when you're involved in physical games, the mind doesn't exist.

Settlement leads to neglect because certainty sends us to sleep (gives us a sense of safety). pioneering leads to attention because it wakes us up (fear of the unknown puts us on our guard/ makes us paranoid i.e. fear being discovered and wiped out (Stranger in a strange land), which puts us on high alert (feel uncomfortable - that is, tense instead of relaxed and happy).

Cooling leads to in-formation (in-vestige-gation) as heating leads to outer-formation (out-pouring's of activity).

Exercise (activity) generates confidence as thought (sensing/ stillness) leads to caution.

Accidents are discoveries, where we panic and run away because we do not know what they are (persistence discloses the truth i.e. going back and looking/ interacting again). We think something is wrong when we don't understand it.

Only interaction cures us (acknowledges that we're bumblers (learners) not experts (teachers) in some area).

When we first learn to do something we repeat it over and over again, to show to ourselves that it was no fluke and how happy we are with ourselves for doing it (easy-peasy). When we don't know how to do something, we not only hide our shame but avoid interaction at all or as often as possible with the thing/ being we hate because it shows us up as incompetent and impotent, when it comes to solving it as a problem/ learning curve.

Sound stimulates us into action (music to our ears/ positive news/ expansive consciousness). Sight/ insight stops us in our tracks i.e. silences us (shrinks our attention down and in, to the microcosmic world of detail).

Time is important for health and social progress. Bodily processes, when too fast, lead to leeching of materials through diarrhoea or if too slow to depositional problems (constipation). In society certain chemical and atomic (energetic) activity can only occur at a certain rate and again rushed or slowed down too much, the inevitable result is failure and the belief that the processes didn't work, rather than the realization that they will only work (exist) within certain parameters: It's more obvious with temperature but less obvious with time. Likewise, the faster you go (beyond your ability to cope), the more mistakes you make (missed connections/ unfinished processes). The slower you go, the less gets done because you don't care about what happens and so put little effort, if any at all, into it (Stress/ apathy ratio).

When you think you can do something about a situation, you tend to panic if things go wrong. If you think you're powerless before powers beyond your control, you tend to sit back and enjoy them (non-resistance versus resistance).

When we surrender to forces beyond our control, we learn from them and survive them. When we resist them, it's a mutually destructive pact.

There are two states of being - standing back to learn (passive observation) and stepping forward to alter (active participation): This is the Artist Effect.

We are most spontaneous when we first give into our desires. It is only with time that we censor our actions and try to control what we and others do (Start fearing instead of enjoying).

17

Observations from another Planet

Life is voluntary - death is obligatory

Fear acts as a brake on our activities, causing us to dig in our heels. Boredom leads us to abandon our hold on where we are and move on to other things/ places/ relationships.

When we try to force others to do things, they resist. When we encourage them to act but allow free will, they are more likely to do so voluntarily. This is because free will is adult (shows trust) and slavery is childish (distrustful).

Co-operation happens where people can see it is in their best interest , to work together with others. Slavery (addiction) occurs when people want an advantage over others (Forces people into consciousness prisons, rather than lets them out to explore reality as free individuals).

The future has no identity - it is turning our attention upwards and outwards, abandoning all prejudices and preconceptions. The past is nothing but identity (Know your place/ Know who you are/ Don't explore (change) - the huddled masses rather than the courageous individual).

The more you resist something, the worse prepared you are, when it eventually arrives (Not prepared at all because you deny its existence).

The root cause of violence is not, not caring enough but caring too much (impatient).

Consciousness is visual (concentration upon) as unconsciousness is sonic (dispersal from).

Observations from another Planet

The illusion of reality is caused by 'movement with' something (proximity). Separation from it allows us us to see it as not being real (distant).

Time is spiritual wealth. Without it we have the panic, which skims through life, making it shallower and more violent (No time to think or space to act, according to our limited thought processes and perceptions). Time builds, fear destroys (fear, interferes - courage leaves alone): Gentleness is strength because it trusts the universe - violence is weakness because it trusts no-one, not even itself).

We have two choices in life when confronted with something we don't like - accept it and change our prejudices (adapt to changed conditions) or move on and take our prejudices with us (reject things as they are).

If you play with children, they'll think life is a game - if you fight with them, they'll think it's a war.

The strong are in a state of repose (calm, sure of themselves, open, honest, accepting of others and whatever life throws at them i.e. present/ centred). The weak are fearful, unsure of themselves, closed off, distant, irresponsible, rebellious, destructive, not wanting to be tied down by anything or anyone.

Those who want to hide something will run off with it, be it an internal memory or idea (thought) or an external object or entity. Those who are open and honest, have the courage to hide nothing, hold onto nothing.

Only doubters learn because they don't believe that they know anything or everything. The arrogant won't shift their position because they think, they know all the answers, so don't look for or at new answers.

Intellectual superiority is where you use knowledge to hurt others (make them feel inferior), just as others use power or wealth as a weapon, to the same negative effect.

All criticism leads to defense (withdrawal/ stopping) as all praise leads to openness (forward motion/ go signal).

When we're in a game playing state of intense concentration, searching for answers, we hate to be distracted by outside considerations or forces. This includes inanimate objects we blunder into because our senses are centred down and in or external noise/ activities created by the outside world - all of which draw attention out as opposed to in. This leads to temper tantrums as we rebel against all that that takes us away from the loved (pursued) object.

Pain is from internal or external expansion (growth), pushing at your boundaries. Pleasure (relief/ release) comes from pressure applied in the opposite direction (external pressure can squeeze life out of you as internal pressure build up, can cause the pain of bloating. Balance is redressing this through movement in the opposite direction (inflation/ deflation (breathing), through the pivot (zero)point of here and now.

By digging our heels in defensively, we can end up staying somewhere and suffering, rather than leaving and possibly ending somewhere better or at least different (A change is as good as a rest).

All emotion is a sign of addiction - the highs of success, the lows of failure; the pain of withdrawal; the fear and anger at the thought of the loss of the addiction; the urge to control the 'loved one' (envy/ jealousy).

Addiction follows several distinct phases, whether it is the individual or society that is affected: Fear that you're not going to get what you want - bewilderment when you do. Calm certainty, when it becomes obvious to you that it is yours for the taking. Excitement at what what you've got, followed by the demand for more ('I want my rights (addiction) fed!'). Then comes the drunk on power stage (achievement), with the carelessness (abandonment of control) that causes accidents and eventual boredom. Lastly comes anger and cold turkey (never again) as you wake up to the results of this overconfidence (the hangover stage). Pain is disconnection at this point (withdrawal - forgetting and allowing the cycle to start all over again: Groundhog Day/ reincarnation).

Anger at our own ignorance (which is a time thing), means we're always running away from our problems rather than staying and facing them (dealing with them). Spirituality is seeing that nothing really matters i.e. we have eternity to think things through and infinity to act within - whereas materialism (addiction) is panicking because you believe something is important, here and now (Feeling you have no room to act in nor time to think your problems through - this is the enemy that dissolves everything (our own fear of failure/ taking things too seriously) as that which allows us the peace to grow and build in, is our friend).

Youth and age are both about distance - the former is moving towards a central point and the latter away from it: Middle age is being there.

Unconscious movement is the release of energy as conscious action is the suppression of it.

The hyperactive (frightened/ excited) need to be grounded, to connect with the real world (sink down or rise up to this level), to see things as they really are - not as they fear (past) or imagine (future) they are: Reality is neutral - it is us that makes it greater or smaller than it is, was or will be, through our positivism or negativity (build things up or knock them down: Optimists make, preserve and rescue as pessimists sever all links to reality because they don't want to be here, now (take response-ability for their own existence).

Cautious people have inner life (thought) - the overconfident have outer life (Are physically active).

An expert is someone whose abilities and workmanship is better than your own, so you admire them for their magical, godlike qualities. An amateur is someone whose dismal efforts you can equal or surpass and who therefore you look down upon as inferior (no secrets or abilities you admire and want for yourself).

The bizarre thing about abuse is that people degrade themselves, in order to be liked (accepted) by others. When they get fed up with the situation they rebel, claiming back their sovereignty ('Do you want to be in my gang?' 'No'). This false inferiority/ superiority divide, requires self-debasement and neglect of your own life, in an attempt to enhance that of others.

Bored adults seek excitement by letting down their guards and letting the young in. The children in return entertain the adults, including self-harm to get their parents attention when it drifts off onto other things (Thou shalt have no other God before me). This is true of individuals in families and class/ sex in society (The criminal and sexual underclasses i.e. workers/ males).

All relationships are abusive because negativity keeps people together through conflict as positivism allows them to drift apart, amicably.

When we see others as islands of hope in a sea of despair, we swim towards them and worship them as gods (or are worshipped as such, if the role is reversed). Upon arrival the distant illusion is replaced by everyday reality and we see them (or are seen) as just ordinary people like us (mere mortals). After overexposure our perception changes again and this home becomes a prison we want to escape from and gods turned mortals become devils, whose very presence is a blight on our lives as we are on theirs: The addiction or relationship comes full circle and dies (The empty petrol tank, having been filled with experience, tells us it's time to leave and continue our journey elsewhere). Fear and awe turns to boredom, then anger and regret as the whole gamut of emotions is played out in this game of discovery and abandonment (charades/ hide and seek/ snakes and ladders). Life is a dream that turns to reality, sours into nightmare, then is gone....

If you're willing to face the possibility of something occurring, then you are less likely to have to deal with it in reality because you've acknowledged the situation. However resist it and it is more likely to sneak up you unannounced and smack you in the teeth for ignoring its existence.

When we're frightened of the outside world, we shut ourselves in and others out (become voluntary prisoners). When we see the world as more full of opportunities than dangers, nothing can contain our enthusiasm for exploring the life (fear/ curiosity ratio swings in favour of external movement as opposed to internal movement or thought over action).

Heat makes it hard to concentrate upon reality as a crisp, clear entity and therefore form cold, hard, certain facts in your mind (memories): Cold is the opposite as it doesn't radiate energy but condenses it into matter (fusion as opposed to fission).

Cold (fear) prompts you into action - heat slows you down as a mechanism (solidification versus melting - 'over-colding' reaction versus overheating (shivering/ perspiration).

To simplify material, we need to break it down into smaller sections and subsections. Things only seem complex and difficult to deal with, when we view the generalized whole (overwhelming/ immense): Every journey begins with a single step.

Verbal language is social. That means it is about assumptions about ownership (pecking order) and asking, not taking. Physical exchange is visual (masculine) and acts without thought for others, just your own urges. It sees 'things' not 'beings' and is the predator, not the herbivore or social animal (ant, bee, termite etc). It is the individual warrior, not the soldier in an army. It cannot talk but it can see. It is the unpredictable outsider, the rebel, the outcast, the hermit, the sorcerer, the sage, the mystic, the insane (unstable) when compared to the rest of society. It is the visionary lost in the future (prophet), not the valued follower of society, living in the present but grounded in the past.

The unconscious is a blank canvas, an empty screen upon which we project our future hopes and past memories. Consciousness is a distraction filled environment, of constant motion / constant calls for our attention, which turns us out into the world, rather than down and in, into the subjective world of creativity, insight and memory.

Consciousness is being aware of where you are and when you are, not mistaking what you're sensing for somewhere else, some other time i.e. superimposing a memory over what you think you are perceiving as being present, here and now.

I've noticed in me and other people that when in an unconscious state i.e. being asleep, we tend to blame the wrong cause, after an incident that wakes us. By this I mean we think that was is there now created the situation which woke us, when in fact it can be long gone or distant.

If you're not aware, you're not present and if you're not present in mind and body, you cannot control your own life or the world around you (Perception is about stopping and 'being' in one place – that is concentrating upon something specific_in the present as movement blurs perception of the present and the world around you).

Consciousness is turning your attention back and down into the self (learning / humility) as unconsciousness is turning your attention up and out into the world or towards 'the other' (pride, confidence, action).

When we 'know' something precisely, it seems to be a purely visual skill of seeing it in detail (no inner voice telling you what is or isn't true, just an outer eye seeing).

If you concentrate on what you're doing, you can see where to stop. If you don't concentrate upon what you're doing, you keep going: This displays the fact that consciousness is actually control of your life and unconsciousness, continuity of movement through space and across time.

Consciousness is actually a defense against unconsciousness (the unknown, the uncontrolled – pure energy as opposed to form, future as opposed to past).

When we concentrate upon the world, time slows down because we take in more detail than normal. When our attention is dispersed, time passes quickly because we only pick up general awareness of something (skip over it, not wade through it laboriously).

Consciousness is what drives society forward. It is the new discovery as opposed to the old problem – without it societal evolution (change) could not occur.

Conscience is a survival mechanism - why have it if it is self-defeating? To be conscious, is to be aware and conscience is telling you what you're aware of, plus how to deal with it. The less conscience you have, the less conscious you are and the more 'unknowingly' self-destructive your acts are within the bigger picture of existence. If you cannot understand the world, you cannot control it and if you cannot control it, you become a victim of your own failure to connect and work with it - dying as an individual and dying out as a civilization or race.

Conscience is a survival mechanism - why have it if it is self-defeating? To be conscious, is to be aware and conscience is telling you what you're aware of, plus how to deal with it. The less conscience you have, the less conscious you are and the more 'unknowingly' self-destructive your acts are within the bigger picture of existence. If you cannot understand the world, you cannot control it and if you cannot control it, you become a victim of your own failure to connect and work with it - dying as an individual and dying out as a civilization or race.

A corrupt society doesn't want the best people employed by it. This is because thorough people will solve problems and that means digging deep in the dirt, to get to the root of why something isn't working, hence the hatred of whistle-blowers and protesters. Lazy and cowardly employees (the ignorant) are already corrupt and therefore will only touch the surface of difficulties i.e. treat them cosmetically.

Force is trying to take a short cut, passed the route of understanding. It is kicking open a door, rather than trying the handle because you don't think you've got the time to figure things out. Criminals use force – the law abiding use patience. Speed is therefore an addiction and tolerance, cold turkey.

I believe that as we get older and more disillusioned by life, we slip more into a depressive state too. This leads to slowed recall. Also, when we are young and enthusiastic, we tend to try to cram as much into our lives / minds as possible but apathy and depression, leads to letting things slide away from us and become chaotic and disorderly. This mess of apathy outside, reflects the abandonment within too or chaos on the outside and mental confusion within. Enthusiasm for life is reflected back at us in detail as depression is mirrored back in the form of 'giant despair' as Bunyan put it, in Pilgrim's progress. When we love life, we enjoy our ability to distinguish one thing from another as opposed to them all blending into one another, when get bored or worse still, despair of them never changing for the better (if we have nothing to look forward to, then it is only natural that we will slide back into nostalgia for better times, when we weren't sidelined but appreciated for being of worth to others. If life is a journey, then the old are heading for the end of it and the young just starting out.

Observations from another Planet

I further believe the slowed, muddled headed reaction of the old is understandable and while the young are waking to the summer of their lives, the old are dropping off to sleep at the approaching winter of theirs. The physical effects of this abandonment of response-ability as the old slide into a second childhood, should not be mistaken for the cause (You wouldn't think a neighbours scruffy back yard's cause was physical, if you knew they just didn't give a damn, would you? So why think the physical decay within is anything but a sign of morale and moral abandonment either?).

I see in the world that life is full of 'creative crusts' or protective containers. The planet has an atmosphere or life couldn't exist and we see it in eggs, seeds, pupa*, mollusc shells, cysts, crustaceans, arachnids, insects, cells with their semi-permeable membranes (countries and their borders), skin, skulls, wombs, factories & assembly lines, cars, planes, ships, submarines, houses, (caves & mud huts in the past, where tool making could be carried out in peace), fridges and freezers, showers and baths, walled gardens, plant pots, tins, bottles, jars and other preservative containers, cardboard cartons, cupboards, drawers, shoes, socks, hats, clothing in general as a protective layer. I see these patterns as necessary for all life to have a stable base, free of outside disturbance, in which to develop - whether it is nature or human imitation, building bodies or artifacts. Is this a new idea or one already noted (* If you view society as a chrysalis, then re-organization of it can be seen in the same way, politically, socially etc)?

Prejudice keeps us trapped in time. Courage frees us to step into the future.

Observations from another Planet

Prejudice is believing that what was once true, will always be true. Experiment is testing this assumption continually, knowing that boredom, caused by self-imprisoned belief, will eventually free us to explore again, when fear is replaced by calm and the courage to move on, prompts us into action.

Prejudice is a way station on our journey through life. It is a break we take, when fear of speeding ahead, slows our progress and we question our advance into the future. Courage and answers get us to move on.

Life is a game of react. When we move, we leave something of ourselves behind that weighed us down with conscience and consciousness. To stay is to win (possess). To move is to lose (leave / become dispossessed, a free spirit again).

Understanding is your calm centre (the eye of the storm - the place you feel most at home). What you don't understand, you fear and want to put distance between you and it (run from that which unsettles you - turns you into a wanderer, a vagrant (the storm itself)).

To understand, we need to be outside looking in. To control is to be in the driving seat (inside, looking out).

Violent people are in pain because they can't understand something and want to reject it, by smashing all links, all reminders of its existence, creating distance between them and it (space / time), so that they can look at it objectively again at some point in the future and gain perspective on it.

Observations from another Planet

What we're afraid of we want to destroy, so it is no longer there to frighten us. If we stop and face it, we neutralize it by breaking it on the rocks of our indifference.

What we hate (fear / don't understand) , we become obsessed by, until its existence overwhelms us, to the point that we notice little else, neglecting our well being and relationships with the rest of the world because of our shame that it bested us in a fight that, had we both won, would have become a collaborative effort (co-operation, not conflict).

Only when we accept loss (defeat / shattered pride) are we free to move on and live again as this is death of the ego or the body.

If we see something as an insult it's because we don't want to be associated with it as it has personal meaning for us that we'd rather not be reminded of (automatic, paranoid defense system, kicking in). If we interpret it as a joke, it's because we see it as meaningless / harmless (non-embarrassing - that is we don't fear it as disclosure).

Intelligence is the ability to differentiate between one thing and another. Lack of intelligence manifests in the inability to make such distinctions.

High morale goes with high intelligence because it stimulates the urge to find out as low morale goes with the opposite (no curiosity / no urge to explore - that is apathy, depression or outright fear of going beyond your limits i.e. feeling / being aware / being responsible).

Observations from another Planet

Death for the mind or spirit, is when you cannot imagine the future, don't want to remember the past and don't want to be aware in the present.

There is a sliding scale between life and death, that includes motion for the body, emotion (energy) for the spirit and knowledge for the mind. The more alive you are, the more stimulated into thought and action you are and the more energy you generate, personally or as a society.

If you cannot face the truth about yourself, how are you going to recognize it in others? The growth of intelligence comes with the subsidence of fear, when you stop and observe rather than run and hide (Avoid contact with reality).

The more intelligent you are, the more open you are to experience new things. The less intelligent you are, the more guarded you are against learning anything new (shutting out change / hostile to the different - forceful and opinionated, rather than relaxed in the company of that which challenges us to think and feel differently to how we did in the past - that is our automatic response to the new).

Ignorance is brutally simple and simply brutal because it breaks things down, not builds them up. A stupid nation is a violent nation. An intelligent nation, like an intelligent person, is peaceful and creative – inquisitive not acquisitive – hungry for knowledge, not hungry for possessions (deep, not shallow). Intelligence is about loving life – ignorance, about hating it (presence and absence/ Alice Through The Looking Glass and running, to stay in the same spot (everything changes): The Tao and the journey in and the journey out or seize the mystery / let go of the answer (bored with the experience)).

Free will has no army because it is led by the heart, not the mind (doesn't try to force things together or push them apart i.e. doesn't moralize – that is think in terms of should or shouldn't)

Science doesn't lead to arguments because it is precise. If water boils at a certain temperature, there's nothing to fight over because it is a provable fact. If however, we say something is hot or cold, this is a generalized statement and can be argued over because it is personal opinion or position that somebody else may dispute, based on their own judgement or feelings about the subject.

In an argument, an egotistical man demolishes his opponent, not his points. Truth suffers in this way – a victim of verbal violence or innocence betrayed by fear (suppression not expression). Blame culture or the Cult of Personality as I use the term, means finding a victim to dump responsibility upon (useless me / villainous you), rather than seek out a mechanical answer to what went wrong i.e. Too little or too much of something, leading to collapse and disaster in some area. Science is the home of the detective, the man finding clues as to what happened, not trying to pin the blame on someone, in an emotional war of fault finding.

Science is about uncovering the truth, not covering up what we fear we might find or indeed are horrified at what we have found out about the world that has thrown our sense of certainty out the window. It is being open to all beliefs but putting them to the test where possible, to find out what is certain, probable totally unfounded or still an unknown quantity as with verdicts in law.

True laws are universally applicable. This means they are not personal but neutral (cover all without fear or favour), so that paranoia and neurotic reactions to their findings are inappropriate and therefore baffling to the viewer of these defensive actions, except as proof they exist.

Why do we need science? Because we hypnotize ourselves and have others hypnotize us into believing something is true, when in reality it is not (advertising and the power of belief to influence the mind). Belief can move mountains but it can become a mountain itself – stubborn, immobile, unchanging – afraid to move on. The truth is an explosion that can bring us back down to Earth again, from our lofty towers of certainty; onto a level playing field of new experiences, new beginnings, new beliefs.

The law of evidence states that the closer you are to something and the stiller you are, the more detail you will pick up about it. It also states that the faster you go, the more distant you are in relation to the subject in time and space, possibly to the exclusion of noticing its existence at all or only having a general awareness of it as a phenomena / object, so we speculate about it, rather than know through direct experience.

Life is an experiment where we posit a theory about reality and test it through experience, each and every day. If we're wrong, we have to adjust our beliefs and actions accordingly. If we're right, it reinforces our prejudices, time however can change the rules (Agar jelly crystallization for instance or dinosaurs ruling the Earth).

We are all scientists / detectives in that we are seeking the truth about events outside our control or that happened before our time. In our alternative mode, we are villains / criminals trying to hide our actions, our motives, our thoughts from others, who might not approve of them (our laxity and its result).

A good teacher does not make others look small (humiliate them)

A good teacher laughs with you, not at you as he sees his own early stumbles in your present ones, rather than tries to hide them, so he appears superior to you now (ashamed of his past)

A good teacher tries to draw out of you what you know, what you can discover or what you can do now: He does not try to overwhelm you with his own knowledge or abilities

A good teacher doesn't rest on his own laurels (isn't lazy or negative)but pushes himself to learn new things as well as encourages others to do likewise. He encourages you to do your best. He doesn't discourage you from trying, seeing you as a rival for his crown. He doesn't encourage others to bully you or do so himself (controls himself, not others).

He doesn't believe in elitism but equality of souls, all struggling to get things right, not perfectionists lost in competition with others but found in self-discovery (aware of what they got wrong, so they can go on to get it right - not stuck in shame of failure in the past but joy of discovery in the present, releasing you into the future through more effort)

He is quiet, patient and tolerant with his students, wanting them to enjoy what he enjoys - the discovery of new lands, new islands of hope

A good teacher doesn't complain that you got something wrong, he explains why you got it wrong and compliments you on trying, encouraging you to think about it more, based on the new information / skill he's given you (Mr Miyagi or Bruce Lee with his pupils in real life)

A good teacher emphasizes future effort, not past failure. He encourages independent thought but joint effort, wherever appropriate, to complete a project (original ideas come from within but co-operation is needed for big dreams in the outside world, to become reality)

He goes slow, to demonstrate what to do, so that the student can see in detail how to replicate the effect and talks in depth, so they understand the subject fully, repeating the material as often as necessary to ensure the lesson sinks in. He does not rush through and skip over material on a once only basis.

Revolt is having too much (glut) as hunger is not having enough (famine)

Observations from another Planet

Consciousness is stopping and examining what you've done. Therefore motion is unconscious / unconsciousness (transmission rather than reception). Motion equals self and consciousness equals lack of motion / emotion (stillness and silence) - a state where we stop 'being' ourselves and become aware of others / other things in our environment instead.

When we lose interest in life (Don't care about our survival/Feel unrequited love), we let go of our hold on life and drift away into unconsciousness (Our conscience dies too for this reason because apathy takes over)

Memory is like a rowing boat, tied by a short lead to keep it close to you, if it's good and on a long rope, if you allow your attention and interest to drift away from the present i.e. the here and now

The faster you go, the more detail is lost to your perception. The slower you go, the more detail you can see. Likewise speed requires greater and greater control at a minute level, to ensure unforeseen circumstances wake us to dangerous possibilities (Highly reactive movement, to avoid collisions).

The frightened want uniformity and conformity. The courageous want dangerous difference

Understanding disarms the violence that despair creates

If you think a problem is insoluble, you're more likely to give into despair and turn to violent expression, in frustration at your ignorance keeping you out. If you think it is soluble, then you're more likely to keep your temper and get on with physically sorting it out, having found a way in.

Science is about studying your fears. It doesn't take the easy way out, by destroying what it doesn't understand (Understanding disarms conflict, by disavowing you of the attitude that you are right, by showing you where you may be wrong i.e. giving you a new perspective / providing you with new or different data)

If you feel you're under attack, than you're more likely to withdraw into your shell and do nothing but brood. If you feel free to act, then you're more likely to burst into spontaneous action, instead of trying to defend yourself from physical or verbal abuse.

If you feel under pressure, time-wise, then you are more likely to ask for help from others, that you feel are more qualified to act than you. If however you feel that you have plenty of time to think and act for yourself (explore / experiment), then you are less likely to seek outside authority or physical skill.

The trouble with outside authority is that it is distant in relation to time and space as opposed to the self, which is localized sensing in the here and now (instant / specific).

When faced with a situation, you have two choices – to moan about it, hoping somebody else will do something about it or take charge i.e. shut your mouth, open your eyes and do something yourself

Observations from another Planet

Anger storms off because it refuses to stay and face something that it doesn't want to be responsible for (abandonment issues). Calm stays and learns (perceives), so that it can control the situation it faces i.e. stays humble / non-existent in the face of pressure, rather than take things personally and run off with hurt pride (feelings of failure).

Intolerance and impatience is a revolt against time. It is refusing to accept that you need to search / research areas you've abandoned, thinking you'll never need to go there again (The humiliation of admitting you're wrong & the need to return to the past, to correct the present, so that you can head into the future once more).

Bound energy is grounded in everyday reality (Solid certainty). Free energy is unused potential (fantasy, (fiction at best) not fact because it has not been turned into reality in the here and now)

When we arrive somewhere new, we're helpful because we want to be accepted (belong). When we want to leave, we become rebellious because we're bored (Do want versus don't want attitude or motivation for action / inaction)

The journey in and the journey out are both stressful because they are rushed: Being there isn't (Arrival i.e. restful / relaxed state of stasis)

Conflict and doubt make a journey seem longer because it breaks up the flow (Shatters the continuity into bits).

We sleep through the journey but awake on arrival, becoming aware of who and what we are, and what we've done to get here (Gain consciousness and lose our unconscious state).

Obsession keeps us trapped in one place, diligently pursuing one goal, until our efforts are recognized. Freedom is having nothing in particular to do, so wandering off wherever our fancy takes us (The free have nothing special to do – slaves are always working on something 'important')

Speed of thought is a talent, not a sign of intelligence, which is the process of learning and understanding

When we find a solution to a problem, we return to flow (movement/ happiness / attention release or freedom from slavery). Problems snag us in unhappiness (stagnation/purgatory or waiting state).

Intention is the strength that guides us on our way – doubt, the brake that makes us stop and reconsider our options.

Intelligence is the tortoise, slowing things down to observe how they work (science/Newton personified) as physical skill is the hare or speeded up action as knowledge develops: The mirror response to thought (introspection) is action (externalization). This is the science behind both these states.

Speed is a sign of physical skill but intelligence, which is based on silence and stillness (observation) is its complete opposite. Speed destroys memory because it destroys presence (not here to sense or make sense of the world). If you are not present to control the effects of something, then chaos reigns. If you are, then order does.

Until you face what you've done wrong,* you won't be able to put it right (mea culpa – I am to blame for my predicament, no-one else as I volunteered for the experience)

*In your own eyes or an act you've seen done by others, that transgresses your own moral code of what is right or wrong.

When we are weighed down by obligations, time slows down for us. When we are freed from them, it starts up again/increases in speed

Extroverts are not violent because their attention is dissipated (turned towards the outer world). Introverts are more violent because they're frightened of losing what they have (become territorial).

There are two types of people in the world – the genuine, who are the same all the time and liars, spies and conmen, who imitate the former to get what they want out of life, only becoming themselves when they let their guard down (stop acting).

Slavery is fear driven action. It is done without attention to detail or even reality because you are afraid of consequences upon yourself. Therefore it is rushed and lacks quality. If you love what you are doing, you take your time and do it in depth (not charging into the future but staying centred in the present: Sensing the here and now, not the feared there and then).

In a crisis, we have two choices – to blame somebody or something else, for what has gone wrong (the past) or try to find a solution in the present, so that things will go right in the future.

There are two types of people in the world – those that use ruthlessness and cunning to get what they think they want, knocking/pulling down those they see as rivals to their ambitions (Last man standing) and those who raise (build) themselves up, through effort and talent because the only rival they see to their future selves, is their own laziness and cowardice sabotaging their efforts.

There are two types of search. One where you sit still and search your memory, for what you've lost, via your last vision of it and one where you frantically search the environment around you, hoping to accidentally stumble upon the mislaid item.

There are two types of emergency – physical, putting life and limb in danger and emotional, where you rush to cover up something that you're ashamed of

Seriousness is a defense against opening up and revealing the truth. Humour is acknowledging what that hidden secret was (no game playing/no hiding the hand life has given you, just total honest revelation at your hopelessness and helplessness before forces you cannot hope to control).

Observations from another Planet

To kill (active reaction) or to let die (passive reaction) is to refuse to be response-able for

Looking forward to the future, our tasks seem gigantic. Looking back on the past, they seem small

The dead know everything but can do nothing. The living know nothing but can do everything (learn / experience / gain knowledge)

When somebody says that they can't afford something, what they really mean is that they are not willing to pay for it

The trouble with the ignorant is that they don't know that they don't know (Are not aware that they're not aware of the truth)

I don't know if this would be of interest to you but I have an idea for an experiment, to show that people perceive things beyond the normal sensory experience. It involves participants with phobias, who sleep overnight in rooms with two way mirrors that have phobic items brought up to these mirrors at various times and their reactions monitored. They won't know where or when these objects or lifeforms are close by because they are hidden and they should be measures at various stages of consciousness or dreaming.

Another observation I had, was the phenomena of needing to go to the toilet. I've noticed that if I'm in desperate need of urination and I'm neither close to my home or a public convenience, then I can only resist the urge to go by staying in a mentally dissipated state. If however I start to project my mind to my final destination, the urge becomes almost irresistible. I'm sure this could be turned into research also, either as an experiment or simple survey because it could be argued that this shows we are more than bodies.

Memory is based either on strong emotion or repetition of effect. The latter works by being predictive (I did it yesterday and I am doing it again today). It is like a hammer drill, wearing a hole in reality, blow by blow, as opposed to a sledge hammer knocking down a wall (trauma) or conversely building one (pride in achievement).

It is also based upon the ability to distinguish one thing from another (the devil is in the detail). It is putting effort in to learn, in order to achieve as opposed to depression, which is withdrawal from the world (abandonment of perception, memory and interaction with the world).

Routine (habit) establishes order, which reinforces memory. Chaos and confusion reign where there is no rhythm to life (no pattern to follow which establishes where everything is and what it does). Interruptions to thought (noise) breaks concentration, ensuring the connection to physical reality is destroyed, turning our attention towards the outside world and away from the inner one, where memory resides (Speed of movement blurs perception, leading to an inability to sense or make sense of reality).

Fear or excitement creates shallow perception therefore as it stops us putting our attention upon the present moment and particular place in time. Intellect is being 'at' some point, whereas positive (e)motion is going towards it (attraction) and negative (e)motion is withdrawal from it (repulsion).

Observations from another Planet

Heightened morale (positive emotion), leads to heightened perception and heightened recall (memory). Low morale (negative emotion / depression) leads to poor perception and poor recall. Mystery (failure to understand) creates the latter state as insight (understanding) reverses this situation.

Attitude determines outcome. The more enthusiastic you are about life, the greater detail you will be aware of in it and the more accurate your knowledge of it will be. Depression, because it leads to withdrawal from the world, means that you will have only a general awareness of reality, if that. Interaction with the world manifests on a sliding scale that we call intelligence, with depression at the bottom (interest in nothing / suicidal state) and enthusiasm at the top (interested in everything / strong urge to live).

The only way to overcome fear is through understanding and the only way to achieve this is through honesty (courage)

Action is always on the cutting edge of now. It is optimism - a leap of faith that propels us into the unknown future as thought keeps us trapped in the past, a pessimistic doubter that misses opportunities

If you stand your ground, you'll record everything you see and have an accurate memory. If you panic and run, the past will be a blur and you'll have no memory because you'll have no perception

The future because it is unknown, can only be reacted to, not planned for. It will either draw us to it (fascination with what insights we might gain) or repulse us (cause us to withdraw in fear because of what we might lose). The past can be planned for because we know what to expect (what is needed to deal with it, from previous experience - that is what works).

To be depressed is to feel you have no future and to feel you have no future is to withdraw from the world. To feel you have a purpose in life, is to advance into the world and act (doubt leads to withdrawal - certainty to advance).

When you have nothing to look forward to, that is when you start looking back upon the past (get lost in nostalgia / let the present fall away).

When we are interested in life, we focus on it in all its complexities (details). When we lose interest, we only have vague, general awareness and maybe not even that (inaccuracy in word and deed).

Destinations are graveyards for the enquiring mind

Nothing is as frustrating as failure - nothing is as disappointing as success

Confidence is feeling wanted (having a purpose in life or at least an enjoyment of it). It is a state of certainty, where you build connections rather than destroy them (suicide) or let them fall apart (apathy). Lack of confidence is your world falling apart or never being built in the first place.

Observations from another Planet

Prejudice is destroyed by action (facing your fears, instead of denying them through violence in word or deed), just as separation is destroyed by physical contact.

Panickers jump into action to hide the truth from others, before they can discover it for themselves. The calm don't what others find out about them because they are open and honest

It is much easier to control a process, if you know when you want it to end. If you don't make a note of when you started , you are then forced to to calculate time taken / needed to complete some task as you go along, which destroys accuracy / certainty and creates panic, rather than a calm, controlled atmosphere

If you cannot predict the future, then you cannot control the present

Memory and sensing are about connecting (being there and seeing / gaining insight) as forgetting is about disconnection and letting go (being moved, physically or emotionally or your consciousness being expanded, through sound (music) as you let go off the small present and relax into the larger field of infinity and eternity

If it's not unconditional, it's not love but politics - the wellspring of materialism (paid loyalty, not spiritual generosity as a natural response)

We make life meaningful by investing in it and meaningless by withdrawing all effort from it

Observations from another Planet

Learning is linking (chain of thought). Violent interruptions (war / disasters) sabotage this connection, destroying intelligence

Curiosity leads us into a relationship with the world as fear leads us to withdraw from it

The frightened command - the confident ask

If you feel betrayed, you betray and suicide is a sense of self-betrayal

When we're under stress, we make mistakes. When under none, we don't even make decisions

A tyrant wants obedience - a saint wants understanding

We all play hide and seek with our lives. We hide when we're ashamed and seek the limelight, when we're proud of what we've done

Fear excludes - courage (tolerance & patience) includes

Limitation creates certainty - infinity creates terror (ego versus humility)

Learning can't be rushed but failure can

You cannot force fairness on anybody because force is the opposite of fairness

Courage is a mountain we climb, to see the world. Fear is a cave we descend into, to find ourselves

Observations from another Planet

Torture and bribery never reveal the truth and are only good for reinforcing your prejudice. Truth is voluntary and admitted with pride, even if it isn't received the same way

Anger is impatience, fueled by fear (panic). Patience is calm and untroubled by impermanence (change / loss)

The stupid fear that they will be stopped by the intelligent, when in fact it is their own stupidity that stops them (limited knowledge / suppressive nature)

Fear divides and conquers - courage unites and overcomes

The poor are happy with nothing - the rich are miserable with everything

Politics is gang membership - spirituality is being yourself, without compromise

One man's meat is another man's life form disposal

Selfishness and stupidity go together because they both imply limit, while generosity and intelligence do not

When you understand, you don't get involved. When you're mystified, you're drawn in by the illusion

Slaves react - free men stop and think (defy their programming)

Observations from another Planet

What we become conscious of, we defend ourselves against as a reality

Love expresses its feelings, hate suppresses all

A happy life is a full life - a miserable life is an empty one

Violence is an addiction because you can't let it go, like all addictive habits

Frightened people are mean because they hold onto things (the past) for 'dear' life, frightened that they will lose them. Generous people let things go because they are not afraid of loss, only in gaining something new (future hope)

What is the danger of the insane? They are more likely to jump into unpremeditated action than the sane (knee jerk reaction), precipitating results that could be fatal to them or others around them. What is the value of heroes? They are likely to do the same

Courage is following something through to completion - cowardice is abandoning or avoiding it

When we panic, it is because we do not understand something. When we do, we relax

Belief can move mountains but it cannot make a lie the truth

Observations from another Planet

To be without motion is to be without e-motion (in a meditative state)

Settlers stay and build (accumulate possessions and knowledge). Pioneers abandon and travel light (disperse)

Tyrants hate spontaneity, individuality and a sense of humour

Love unites - hate separates

During peace, everybody matters. In war nobody does

Love gives - hate takes

Positive emotions fill the world - negative ones empty it

Life is the truth - death is the lie

Aging is wear and tear on the body as it moves through time and across space

Anger is impatience and intolerance in action (fear provoked thought)

Life is a trivial pursuit - death, a serious business

Truth is black and white - lies are a grey area

When we are honest about our feelings, we react. When we lie, we rehearse / plan our public actions

Observations from another Planet

Things start with attitude and end with gratitude

Death is life's way of telling you that you've outstayed your welcome

Individuals have the courage to stand alone - cowards hide in groups

Feed the ego, starve the humility to make them grow

You don't have to kill your enemies to overcome them, just understand them to death

Peace allows diversity to develop - war enforces conformism

Identifying with others disables you as an active helper because you take on their symptoms. This helps understanding but leaves you as helpless as they are

Lies (suppression / paranoid defense) slows things down. It is fear of the future, the unknown overwhelming you and replacing past certainty (control) with doubt

You have nowhere to go but here, nothing to do but live in the now

To act, speed up. To understand, slow down

Sound gets you started - sight stops you in your tracks

Observations from another Planet

Tension is stored memory (defense), which relaxation releases

Loyalty is who you betray and to whom (changed allegiances)

A suppressive society destroys - an expressive (open one) creates

Anger stirs us into action - fear gets us to stop and think

Change stimulates (wakes us up). Sameness sedates (puts us to sleep)

Life is composed of log jams that require thought plus solutions that require action

Sound stimulates outer action as sight stimulates inner thought

He who questions doesn't act. He who acts doesn't question

Love leads us towards something. Fear keeps us there and boredom releases us

To kill is to destroy somethings potential development

Making a choice leads to action, indecision to inaction
We are the victims of our own actions

Swift action destroys - slow action builds

Observations from another Planet

Openness assuages fear by feeding curiosity with knowledge (experience)

Life is diversity - death is singularity (reductionism)

The highly reactive, poorly analytical jump to the wrong conclusions and act on them. The highly analytical, slowly reactive, don't jump to conclusions and don't act on them as quickly (deep thought as opposed to shallow thought)

When we don't understand how something works, it is terrifying and bewildering magic as we can't control it. When we do understand it and can control it, it becomes science (common sense, common place and therefore easily reproducible as an effect)

A broken mind breaks the universe into bits. A healthy mind welds it into a whole (takes control of life as opposed to fearing it, becoming powerful, instead of overpowered)

To do things the way other people want, is to learn a new way of doing them. To do them the way you're used to, is to do them skillfully however (habitually)

The more sensitive you are, the more incapacitated you are too, through being open and receptive (slowed down or stopped in your tracks as with the old, babies and the learning disabled)

Knowing where you are in relation to where you ultimately want to be, is more important than knowing where you are at a particular point of the journey

Observations from another Planet

Stability is sanity. Insanity is what moves us emotionally, whether it is positive excitement or a negative panic attack

Change is tiring because of the adjustment to the challenge of new circumstances (learning curve). Routine leads to the insomnia effect because we know everything in detail about a situation (bored / orientated to the old)

There are really only two states - thinking without doing and doing without thinking (lost in thought or lost in action / habit or originality / unchanged body or unchanged mind)

The old bores and introverts - the new extroverts and stimulates our attention (old habits resurface as you settle - new experiences distract from them)

People who think they are right, try to force that certainty on others. People who doubt even themselves, discover the truth through observation

We all need a safe haven - without it, we are driven violently insane by constant motion / emotion (no time to ourselves, in order to develop character or depth): Even gangsters seek sanctuary in a church

The growth of individuality comes with the realization that you are projecting your own personality (beliefs / habits) onto other people and that these may not apply to them (We all breathe but we don't all darn our socks: This is general awareness versus specific differentiation i.e. the learning process / seeing / enlightenment)

Observations from another Planet

All illusion is caused by distance - all disillusion by proximity (The perfection of Lilliput versus the overwhelming input of Brobdingnag)

The new disrupts the old because of its energy and enthusiasm. The old corrupts the young through its disdain and boredom

Memory, like perception, is of time and place. If you're not here now, you don't have reminders in real life, visually, to drag abstract associations back into your mind, verbally

Happy people don't interfere with or criticize how other people live their lives because they are too busy enjoying their own

Apathy is simply feeling not wanted and retreating from the real world therefore

People need to recover from trauma (shock / injury). They do this as a natural response through withdrawal, to assess damage and repair losses, whether emotional or physical. When ready to meet the outside world again, they automatically step forward and interact

If you're afraid of being interrupted in what you're doing by external forces, your attention will be aimed outwards, not inwards. This is what causes failure in your endeavours as success requires attention upon the here and now, not the there and then

Observations from another Planet

Drawing attention to someone's bad habits only reinforces them (criticism leads to rebellion). Polite people ignore such behaviour because it is counterproductive to oppose others as they've learned through experience - that is, it leads to bottling up and ingraining of traits, which time would release through the maturing process

People who are not present, are startled by life pulling them back into the world. They repeatedly ask the same questions because they're not listening to the answers

The reason we don't know what we've got until it is gone, is that it's part of a connected whole / memorized map and removal / alteration / change, shatters this holistic / immortal seeming view (damage / death).

Our most dominant memories override more recent ones – hence our first thoughts may not contain the correct information, when compared to reality (the fruit machine of choice / variations of results)

The new frightens / excites us because we don't know how to handle it as the old is safe /bores us because we do know how to deal with it (predictable) – depending upon whether you take up the challenge or reject it (withdraw from contact or advance towards it)

People who are always right, never go beyond their limits. People who get things wrong are willing to explore the unknown, the possible in order to learn something new

Observations from another Planet

Fear never sleeps because it is always on the alert for potential threats to its safety, plugging any holes that might allow danger to leak in and change its world or world view forever

The lazy and the cowardly can always find a reason for not acting, based on past failure but the innocent are always dumbfounded when accused because they haven't rehearsed their excuses, until word perfect (expect the worst from others rather than the best)

The vain want to impress others (care only about appearance). The sensible only care about comfort (how they feel, not how things appear to other – in other words they are not trying to sell their souls, in order to manipulate other people's feelings)

People who neglect things around them are either lazy (demoralized / despairing / apathetic) or stressed (too busy to get everything done in the time available). The latter needs physical help and the former needs emotional support (encouragement / inspiration) to lift their spirits

Happy people are independent and encourage others to be too. Miserable people are dependent (addicted) and encourage others to be the same (demolish their confidence, not build it up)

The small minded sabotage the connections that allow big minds to grow, by continually interrupting the flow of ideas from one link to another

Observations from another Planet

People who are asleep have no memories except of traumatic events, which wake them up because they are loud, unexpected and violent (age is about going back to sleep (hibernation) as youth is about waking up and becoming aware of everything and anything)

A leader is a pioneer, who doesn't want to be where he is but where he isn't. He seeks the unknown in mind or body because he's fed up with the known (the predictable), which he no longer wishes to be a part of, lost in habit, a slave to conformity – that is a follower caught in hope, not self-led – that is released by despair at his situation

The hysterical personality is never at rest. The calm personality continually tries to centre itself (fear wants to control the outside world – courage tries to control itself)

Speed creates stress. The faster you go, the more alert you need to be, to change (avert potential disaster). The slower you go, the less stressed you are as your focus widens (relaxes)

To be centred is to turn inwards and find nothing. To be unbalanced is to turn outwards out of fear, of what we might find within ourselves

Experience distracts our attention from going within, to make sense of the world we face (gaining understanding of the phenomenal world)

People who have no memory have no inner life / light. They are what I call repeaters – asking the same things over and over again because they don't hold onto anything. They have rhythm in their lives because they seek certainty through habit

Phenomena addicts seek continual distraction through action, rather than seeking a calm, fluid world of rhythm and harmony (seeming external monotony / centred reality rather than 'out there' all the time)

Why are we overwhelmed by difficulty? Because problems seem blown out of all proportion (generalized / all over), whereas solutions shrink reality to specifics (small, obvious, simple, insignificant, individual victories rather than giant defeats)

When we feel sorry for ourselves, we get sucked into our own little world because of what we don't understand about the outside world. When we feel sorry for others (have a solution for communal ills), we are pulled back out into the external world

Problems beget solutions. Solutions beget problems. By this I mean when we're on the cusp of a dilemma – it's either fear / excitement (Do I try to solve this problem or abandon my life to it?) or boredom / challenge (Do I stay safe or risk an adventure?)

When we're overwhelmed by the enormity and seeming complexity of what we face, we feel suicidal because things appear beyond our control. When we regain our composure, no matter how small the foothold in this world, we want to live again because we no longer feel helpless victims of our own insignificance (ignorance) but victors over it

Observations from another Planet

We protect ourselves from change (shut out the new, the different) in order to maintain the status quo (sameness) but like breathing, it is inevitable that it will pour into our world anyway

Fear is that which disappears when we face it, like it never existed. When we try to shut it out, it reappears like it had never gone away (was always there and would always be there)

Failure and panic go together, when other people are involved because we fear being exposed, so rush around trying to cover up evidence. When there is only us and God, we have a more relaxed, forgiving attitude to the learning process

A frightened society is a secret society (distrustful) i.e. hides bad news because of the fear, of consequences (negative reaction). A courageous society is an open society (trusting – deals with events, not hides them) and so turns them round from disasters to triumphs: treats things lightly, not as irrevocable failures. All of this applies to individuals too, in family and societal situations. This is how we make the world or fail to make it.

When we are frightened (angry) we cannot see the truth. Only when we calm down does the shame and other negative reactions dissipate enough for the blinkers to come off and the bigger picture becomes to light

Change disrupts and is brought about by dissatisfaction with how things are. It leads to counter disruption, by those who oppose that change because they see it is in their interest to leave things as they are (protestors and the police / rebellion and the army / war and another country's army)

A happy man sees nothing wrong in the world, so floats upwards to heaven because he doesn't interfere in other people's lives. A miserable man sees nothing right in the world, so sinks down into the depths of despair because he can't leave well alone

Nice people finish last because they are just out for a stroll through life. Nasty people end up in the graveyard, hospital or as the last man standing, holding the worthless booby-prize.

When we're obsessed, we're reluctant to let go in case we miss something. When we're bored or depressed by something, we let go of it willingly and easily (open up to change instead of defend yourself against it)

We impose our will on others because we fear (imagine) what they will do with their own free will. This distrust is what drives bullies and dictators, with their pre-emptive strikes (unpredictable behaviour) as opposed to the predictability of those who want peace or try to avoid war

When we are repulsed by something, we let go all connection to it and it is this that makes us ill. If we enjoy every experience we have, we stay connected to the world, no matter what it throws at us (Asterix The Gaul and The Big Fight, where Getafix and Freudix carry our potion experiments on themselves, laughing hysterically at the results)

The faster you go, the more you hone down your lifestyle. The slower you go, the more you build up your resources and knowledge - that is you become grounded in reality, which means in full control of your faculties and environment: To be ungrounded, is to be light headed and traveling light (no thoughts to fill your head, just wanderlust and curiosity to draw your feet out there into the world, rather than in here (the mind)).

To change is to learn. If you don't change, you don't learn anything new about events in your life (refine your actions, from the crude knowledge you start off with - that is the initial discoveries you make)

When we are heading downhill in our tasks, we get intoxicated at the thought of finally finishing. Likewise we get angry when a new problem throws a spanner in the works, of our sense of perfection (completion).

Paranoia is believing everything is down to you (your fault). Reality is waking up and seeing that others / other things exist and happen in this world that you have no control over, freeing you from total guilt and instead filling your mind with knowledge that you are not alone and not responsible for anything in this world but your choice to connect or disconnect from some part of it

To solve a problem, we have to stop ourselves first, in order to observe and learn what makes it a problem for us or the world, then we can start seeking solutions until we find one that works best for us. To ignore it by rushing passed it, is to perpetuate the situation as an irritation, tugging at our consciousness

It is when we're asleep that we think we know but to still fail at a task. It is only when we truly wake up to reality that we see the truth and succeed

When we're trapped in confusion because we believe that we know what we're seeing, we can't solve the problem we're facing (fixed idea). When we realize what we're really seeing - that is when we truly solve the difficulty

When faced with a challenge, we either try to escape it or pursue the connections that lead to its source (learn from the experience)

The righteous accept their fate (feel at home in their bodies and their lives). The wrong ones, rebel and want to escape from what they see as an unfair life, an unfair world

A frightened man will only be interested in what is the quickest way to do something. A calm man will seek out the best way to do it (quantity of effect over quality - short lived or long lived phenomena)

When we settle, our prejudices build up around us as defenses to keep our lives as they are. Anything that moves us, physically or emotionally, changes our perceptions and lifestyles (attitudes); knocking away all preconceptions of how things are or should be (survival instincts cut in and we move on)

When we don't understand how something works, we are more likely to interfere with its running than if we do: If we do understand how it works, we are more likely to be calm about its activity and let it run on, uninterruptedly (fear sabotages, with its ignorance - knowledge pacifies with its revelation)

We're repulsed by what we don't understand - that is we need to step back, to observe and learn from the situation. What we're attracted to, we interact with until bored we're with it (know it in intimate detail - that is it becomes too predictable)

Sometimes we think something is wrong, simply because we don't understand what is right (how it works - its processes, stages, development and activity), so we dismiss it as flawed when it is really our perception of it that is.

Heaven is wherever you feel understood (home). Hell is wherever you feel a victim of circumstances beyond your control (displaced person / refugee / stranger in a stranger land)

If you are content with your life, nobody can tempt you to change it (where you live, who you are, what you have, the job you do, the challenges you face everyday).

Speed lead to a lessened perception of time and therefore less patience and tolerance because you feel starved of it. Stillness however leads to an increased perception of eternity (all the time in the world) because you are not under pressure to complete actions by a certain time period (no tension through action i.e. shrunken perception - just relaxed (expanded) appreciation).

If you can tolerate all the minor nuisances in your life, you can handle them so that they don't explode into major disasters, through abandonment of all control: Intolerance is the violent rebellion led by the ego, you see in bullying and dictatorships.

When you panic everything seems important and disastrous. When you're calm, nothing seems important or worth bothering with

When something goes wrong and you're in a reasonable frame of mind, you ask yourself why didn't it work? When you're in an ignorant mode (panic - no time to think or act), you shout commands at it like, work dammit! Discoveries are made in the former, calm state which arouses curiosity and leads to cover-ups in the latter state, when negative emotions take over (fear, shame etc).

When we rush time shrinks. When we slow down it expands as does our perception of everything else. Panic is shrunken time as calm is expanded time, in which we can include everything instead of exclude more and more. This is the basis of knowledge and ignorance, patience and impatience, tolerance and intolerance – shutting in and shutting out as opposed to simply letting all things be.

Observations from another Planet

You are either happy where you are or miserable because you want to be somewhere else (out shopping, having a meal, watching a film, moving to another house or another town, country, planet or even a different epoch in history, future or past). You're either happy being who you are or again, you want to be someone else (change your name, face, sex, social circle, job, nationality, race, sexual orientation, age, IQ, qualifications, identity in some other way).

You either want acceptance, approval or acknowledgement from others, which makes you their slave or you accept and approve of yourself, needing no acknowledgement from outside yourself (peers, family, church, community, government, world etc.). You are either your own person or nothing.

Life is swimming upstream against the current - towards the future with hope. Death is letting go (apathy) and getting carried back downstream, towards the past.

If you want something to be sacred, you have to treat it with respect because your attitude decides the actions you take and therefore reflects your own mind (it's identity will be your identity, no matter what you say the truth is).

The pretentious are easy to manipulate because they are already fantasizing (lying) about reality. You cannot manipulate truth seekers because they are only interested in the genuine.

Regrets are the nails in the coffins of life. They limit where we can explore, in case we bring up memories of old wounds. Mentally this ages us as innocence leads us to make the blunders that start off the cycle again.

A confused mind cannot remember because it cannot separate and choose between seemingly similar or the same material. A clear mind can see every little detail, which tells it the difference between forms and in time.

Memory is the ability to distinguish between similar items of time, place or person. Forgetfulness is where they are all melt into one. In other words, separation is necessary for remembering as slowing down is for perception.

If you don't identify with something (your appearance / beliefs) then nobody can insult or upset you as you have nothing to defend (no past). Only identification makes you a target that can be hit.

When we see a problem as limited, we are more likely to tackle it. If we see it as limitless, we are more likely to abandon it as an issue (see it as not being resolvable but totally overwhelming instead).

Finding the principle, links effects across a broad spectrum. Not finding it leads to superstition (actions 'just in case') or individual handling of what seems like unrelated issues (monomania as opposed to exploration and discovery).

We torture people, not to extract the truth but to reaffirm our prejudices (get them to confess to to the lie we wish to hear from them) just as we try to suppress the truth, through verbal and physical abuse, in other forms: Truth comes out when we relax, like flowers after a storm or refuge seekers after an air raid - hence in vino veritas.

Tolerance and patience ensure we gain depth of knowledge about a subject before judging it and deciding what to do with it. Impatience and intolerance react to a situation on the flimsiest of evidence and therefore is less likely to be right about their assumptions of cause or guilt.

A belief in the importance of something, forces us into action (rushing). A belief in the unimportance of anything, leaves us totally unmoved, emotionally and physically (not caught up in the games people play).

Why do the ignorant remain ignorant? Because they don't want to know something exists (see it) or be reminded of it, through word association - hence illiteracy.

If you see life as your ally, a world of opportunities beckons - in which to change / be changed (metamorphosis of mind or body). If you see it as your enemy, then it blocks your every move (stagnation / prison / change only for the worse, in your opinion).

A belief in the importance of something, forces us into action (rushing). A belief in the unimportance of anything, leaves us totally unmoved, emotionally and physically (not caught up in the games people play).

Power comes from the release of energy / attention. Powerlessness (or depression) comes from having your attention pulled down and in by a problem or question as it is released by an answer or solution.

Patience creates everything by doing nothing (allows the time and space necessary for things to grow). Impatience destroys everything by trying to do everything (allows no time or space for anything to grow (peace in which to evolve (change) as an individual, society or planet).

If you're present in the world (inside, looking out) then you perceive it as it is and remember it as it was. If you're journeying towards it, you imagine it and if you're leaving it, you forget it (lose consciousness of the world and your place in it)

We can never know what we're doing, only what we've done (completion brings understanding). Life is a series of blunders and death (the end of something), which is the illusion of perfection.

When we shut something out in fear, we can never be sure what it is. Only by letting it in (sensing it) can we be sure of its identity: This is the whole basis of knowledge (experience) and ignorance.

When we feel unappreciated (bad about ourselves / negative), we want to leave where we are and cut ourselves off from others. When we do feel appreciated (good about ourselves / positive), we want to stay and build upon our relationships

Observations from another Planet

We wound others by word or deed because we want them to withdraw from the world, so we can advance into it or maintain our position here (life as an arena theory).

The courageous forget themselves (become 'other' aware / lose themselves in action). Cowards remember themselves (become self-aware / lose themselves in thought): I use cowards and courageous here as neutral measurements of direction, not emotive terms to encourage or discourage behaviour (prejudice).

When somebody attacks you, you lose faith in them and are tempted to sabotage the relationship further on your side too. You may know it's fear speaking or acting on their part but it still doesn't stop it destroying your morale, built on loyalty.

Those who see only the finite (the past - the end of things) are more likely to be mean than those who see the infinite (the future - the beginning of new things).

When we react to a situation, our inner voice is lost in that action. When we stop, the inner critic kicks in and we hear outer voices too. This is why we deliberately get lost in action as a ploy and accidentally jump into action, in an emergency (no thought, no hesitation - just seeing instantly what to do and doing it: Stopping allows us to assess the results of those actions and inactions plus feel them too (shock, injury, joy, sadness or emotional collapse)).

To take things seriously is to feel envious and inferior, so seek approval / notice from others. To laugh is to not be impressed / depressed by something (not lust after it, to seek balance / poise again).

Seriousness is about stopping things as humour is allowing things to continue (seeing them as harmless, not harmful and so not defending yourself against them).

Responsibility (courage) allows diversity to exist. Irresponsibility (cowardice) wants uniformity and conformity (simplicity of form and thought) because it fears the new, the different, the unknown.

Success is fluent (flows). Failure chugs along, slowly, stiffly, hesitantly - feeling its way (learning, not confident, competent, certain, accomplished).

To hate is to not understand. To love is to make sense of the world and the situation you find yourself in.

Violence comes from ignoring others - peace from paying attention to them (acknowledging their existence and right to exist.

Strong people are merciful because they know they don't have to be. The weak are merciless because fear drives their suppressive acts.

When we think we know the answer to something, we get stuck in habit (the safety of the ego, the known). This is because change (the new, the unknown, the different) throws our sense of safety and certainty out the window and leaves us feeling vulnerable. Certainty of belief, leaves us feeling smug and snug in our ignorance of the greater truth that we shut out.

You can only deal with reality, if you're willing to face it. Those who care more about appearance than practicality, will hide the truth to avoid upsetting other people (sweep problems under the carpet, where they build up because they are not dealt with).

When we transmit ourselves into the outside world, we empty ourselves of thought or feeling. When we are in a receptive state, we draw in perception and ideas.

To relate to anything (life / existence / another being) is to be calm, relaxed, present. To be tense, nervous, lacking presence in mind, body or spirit is to lack a relationship with anything (no conscious awareness of / no conscience towards the continued existence of something, including yourself).

Negative people want to make you feel bad as it makes them feel good (inflates their egos). Positive people want to make you feel good as it humbles their egos, to encourage yours.

The start of something is more difficult to deal with than the end because you're not adjusted to the change it brings. Those we consider fit or skilled are acclimatized through experience, to those things they are good at. They are not born better at it than anyone else, just habituated to it more than others.

Observations from another Planet

Joy comes from the release of energy (speed / movement). Misery comes from binding of energy (hitting a brick wall / reaching the top of the glass ceiling). The first makes you feel a success - the second, a failure. Joy shows you the world in general - misery, the world in detail.

We censor our thoughts, feelings and actions to fit in with whatever group (family, society, gang etc) we wish to belong to - that is become typical (conformist), not stand out as atypical (rebel / ourselves).

Anger drives us and is based upon the fear that we don't have enough time to do what we want (constrained). It makes us ignore others, irritable over hold-ups and violent towards the outside world as we take out our frustration on it. Peace is believing we have all the time in the world, to do what we want to (nothing is important / nothing is serious and we have the tolerance and patience, to pay attention to the existence of others / other things. Speed, pushed to the limit, leads to destruction and rebellion (the urge to slow down and examine what we have done / are doing).

A problem is lack of direction (being snagged on something / stuck in limbo / caught in a log jam). The solution is to slow down, analyze what is going on and free your attention from where it is trapped, by disentangling the components, so that flow in every direction is restored and conflict resolved.

When we accept our basic worthlessness, we stop rushing around trying to hide our imperfections, exhausting ourselves in the process and instead settle down into a calm, meditative state.

Those who boast about their past knowledge, verbally, are not as aware of the present moment as those with their eyes open, now, who are perceiving reality as it is, in that instant.

When we react defensively, we shut out experiences and shut in memory that we'd rather not face. This leaves us shallower than if we'd relaxed, opened up and learned what life had to teach us, which would have made us deeper, more thoughtful people

When we are in control of our lives, we find practical solutions for practical problems. When we lose control (get angry at the situation we find ourselves in), we lose our minds (memory / ability to perceive) because our energy (life attention) pours out of ourselves, rather than in.

Skill is tight control, acquired with time, which starts off as loose inability to control anything and gaining speed as experience pulls in attention (fear of failure before an audience turns attention outwards).

Stage one despair is memory loss as you try to forget everything that fills you with repulsion and self-loathing. Stage two is losing perception / connection to the world as it becomes unreal / surreal to you and you become clumsy and blind to what is under your nose and usually visible (but no more).

Observations from another Planet

Laziness is the attempt to cover up the truth. It slows down our reactions and causes us to withdraw from the world, creating misery. Happiness comes from exposing the truth as it speeds up our reactions and connects us to reality again (movement generates energy, through release (heat) as thought creates mass, through slowing energy down (cooling it).

Hatred (prejudice) depends upon memory of the past. Love is in the present (innocence) - no expectation, just experience of what presents itself to you in the here and now.

If you want something, you're willing to stay and tolerate anything to get it. If you don't want it enough (price too high / response too low), you'll react and leave. This is the reality we all face every day. Every living thing has to adapt to changed circumstances or sever connections with this particular space / time continuum (here and now) - that is leave a relationship / die.

The difference between a panicker going at speed and a skilled person, who's worked up to this point from a slow start, is control. The panicker is trying to get away, so is ungrounded - the skilled observer is trying to maintain contact in the here and now/ get closer to the phenomena, so is grounded in the detail of life / existence.

Fear isn't spontaneous. It hides behind rules, regulations, rituals, routines and habits as defense mechanisms. True spirit (courage) flows because it isn't rigid but formless. It holds onto nothing, obeys nothing but its own heart, its own feelings about life and reality - bending and breaking the rules as it sees fit, washing away conformity in a tide of (e)motion.

Discoveries can only be made in the present. This means blaming the past for what it did to you, is avoiding changing things now, so that they are adjusted to your current needs (other people are not your slaves or parents, eternally catering for your wishes. For this reason God gave you individuality (adulthood) so you could look after yourself.

If you want to calm somebody down, including yourself, do nothing / go nowhere - just be there for others / yourself. If you want to excite them into action, do something / go somewhere (move in other words, don't stop and think).

You sometimes have to suffer to gain knowledge because presence is required for experience, which then leads onto understanding / insight but that doesn't mean pain on its own will bring enlightenment.

Hope is what draws us into the future, in the belief our existence can change the world for the better. Despair is where we give up trying (abandon all effort in the present and leave it). A world without hope is a dead world.

Optimism is that which isn't afraid to tackle anything and is never overwhelmed by the amount or size of the tasks facing it (it is a successful juggler). Pessimism sees mole hills as mountains, time as limited and abandons every challenge without ever trying them.

Observations from another Planet

People give reasons for their actions or excuses for failing to act. It's not a question of right or wrong but choice and decisions - that is response-ability and the willingness or unwillingness to act (hidden or open motive). Courts don't decide on morality but glibness of tongue, which is the best excuse or reason for action / inaction, that a lawyer can give. It's all just words versus judgement on actions (belief, not truth - results, not ethics).

When we accept the truth and express it, we move on with our lives. When we lie and obscure the truth, we are trapped in the past, seeking revenge against what we see as unfair.

The reason we repeat actions, unknowingly, is that our attention drifts from the present and on our return we cannot remember having been there in the first place. If we concentrate wholeheartedly upon the present, without wavering, then we have nowhere to return to, no 'time' to forget (always here, always in the now).

Ambition is driven because of feelings, of time being constrained. This leads to impatience and violence, verbal or physical abuse, when pressure is put upon the person / group to slow down or stop what they're doing. Left in peace, people complete their tasks without rebellion (conflict). Worriers stop warriors. Doing always puts you under pressure, not doing doesn't and can't (passivity).

The expectation cycle is where we predict a particular reaction is going to occur, in response to something we're doing now because it has in the past. Because of this we jump into defense mode over it, even if nothing does happen (no negative counter response or criticism). Being pre-programmed it doesn't take note of present circumstances, which may have changed drastically from past ones, that are reliant on memory, not perception.

Only when we stay and confront reality, can we truly, truly see it for what it is and understand it. All movement from this inner journey is avoidance and evasion of this central fact. It is ignorance rather than acceptance, hate rather than love. It is running from, rather than towards existence.

That which frustrates our aims, we resent for stopping us. We loathe it and are physically repulsed by the thought of any contact with it - making us prejudiced against its very existence.

Enlightenment is realizing that everything you thought was self-help, in fact was self-harm. Being an avatar is simply trying to spread that awareness to others.

We tell lies and keep secrets, when we want to maintain or create independence from others. Being open leaves us exposed to manipulation and persuasion by others. It also ensures the status quo is maintained (stability / lack of change).

The trouble with authority is that it leaves the impression there is a right way to do something and only one way to do it. This leaves no room for creativity and originality of thought or action i.e. no belief in the arbitrary nature of reality but one of rigid conformity instead.

Unpleasant associations (horror / disgust) come when perception is slowed down, so that we can see what speed glossed over (happy action): Warts and all becomes visible, where generality misses out details.

The frightened are loyal to their fears (guarded / stay ignorant). The courageous are loyal to possibilities, exploring potential (open to knowledge / experience).

The truly depressed can't reach out to the future, through the medium of the present because they've collapsed back into the past (feel failures / in retreat from a world they cannot face).

When you're depressed everything melts into an amorphous mess. The ability to distinguish detail / difference is a sign of involvement with life and an enjoyment of it, which is the opposite of depression.

The faster you go, the shallower your connection to the world and the cruder the results of your efforts because of this. The slower you go, the more control you have over your life and therefore the more refined / detailed the results and your perception of it.

If you treat things with respect, by slowing to observe and acknowledge their existence, you can avoid unnecessary conflict with them. However if you ignore their existence and charge through them and all over them, to get what you want or where you want to be, then your speed and ignorance will injure you and the only danger is from you attacking others, not their existence (your entanglement in their lives through haste, rather than avoidance through thought).

Murder is denial of life. Rape and burglary are denial of privacy. Overcrowding too stops development of mind and body as all are infringements of the right and ability to grow to your full potential as beings (be alone with our thoughts, to sift through them / have room to expand physically too).

Angry people, full of hate, have definite views / want to control situations. They want to create distance in time and space between them and a situation they're having trouble reconciling themselves to. Calm people have no particular opinion on anything (are not easily moved or disturbed by the thought of particular issues).

What's the point of getting other people to do things for you, when you're quite capable of doing them yourself? It's like getting a surrogate to weight lift and expecting to put on muscles yourself or getting somebody to read a book and thinking you'll retain the knowledge, or enjoy the emotions that reading would have done, had you read it yourself, or even getting someone to sense the world for you and expecting to make sense of it yourself, without any experience or effort - reality just doesn't work that way.

Only those alive in the present moment, can perceive (experience) the truth because that is where it lives - not in the past where memories of what was exist or in the future where imagination of what could be exists.

Observations from another Planet

When we're in our comfort zone, we're lovely to know because we feel safe, certain, knowledgeable. When we're in our discomfort (challenged) zone, we act ugly because we feel vulnerable, unsafe, threatened (in unknown territory / the learning curve).

The new stimulates - the old sedates. This is why all experiences lose their effect over time (become invisible / white out conditions).

Optimists want to raise you up to their level (heaven / positivism). Pessimists want to drag you down to their level (hell / negativity). This is the basis of all relationships, good or bad - conversion to one side or the other / one direction or the other.

Everything we do is a leap of faith because we cannot know something unless we've actually experienced it. All else is advertising or opinion - subjective knowledge, not sensed reality.

Negative people want to control others because they fear those with autonomy. Positive people are just happy being themselves, so have no urge to interfere in the lives of others.

Everything sufferers from the effects of wearing down, through use or clogging up, through being unused (backlog).

Mystery leads to confusion and misery. All answers lead to clarity and elation (chaos replaced by order in mind and body).

Observations from another Planet

Insight is that leap of understanding, when you realize why something doesn't work and invention, when you realize how to fix it.

Problems are either psychologically or physically based. Deciding which is which is half way to solving the puzzle.

If it's not unconditional, it's not love but politics, the wellspring of materialism (paid loyalty, not a natural response to life's generosity).

The only reward you get from politeness, tolerance and patience is peaceful co-existence with the world around you. The reward for greed, impatience and selfishness is short lived material advantage (looking over your shoulder continually, for the revenge you fear will come someday).

Accidents happen when we relax because we release control (turn our attention up and out / forward to an imagined future) as opposed to gain it through concentration (turn our attention down and in / back upon reality).

A joke includes (is understood). An insult excludes (is not understood). Attitude is all.

People who are ashamed of their actions, try to hide them - especially from themselves.

War is what we do (how we react). Peace is what we don't do (how we don't react).

Observations from another Planet

To understand, you must stop and look. To engage, you must listen and act.

Boredom is withdrawal symptom from action (activity as a habit).

To solve a problem, you must stop and examine it. To rush passed is to leave it in place, troubling your conscience / consciousness for eternity.

We can never understand why something has gone wrong, only why it has gone right.

When the mind goes to sleep, the body wakes up (the ego is hid). When the mind wakes up, the body goes to sleep (the id is hid).

Peace allows specialization (diversity) to occur. War (disaster) leads to a breakdown of society, into generalization of skills (uniformity / conformity).

When we're happy to be where we are, we relax. When we don't want to be where we are, we tense up and try to move away.

A servant is happy because he has something to look forward to. A master hates because he has arrived (is disillusioned).

In life we have two choices - to treat it as a tragedy, mourning what we've lost or a comedy, delighting in what we may gain.

Observations from another Planet

To build is a slow, steady process, based on patience. To destroy is fast paced, erratic movements, based on fear and impatience.

Anger disconnects (abandons). Calm connects / stays connected or reconnects (assesses situation / seeks a solution to the problem).

Success in any direction, mental or physical, requires concentration and continued contact, to reach your destination or goal.

Logic is working back from the present, to deduce what happened in the past (cause of effect). Intuition is working forward from the present, to deduce what is heading our way from the future.

The trouble with the blind is that they cannot see that they cannot see (They think the limits of their perception, the confines of their consciousness prison, is all there is: The shadows on Plato's cave).

Consciousness protects us from further harm, by getting us to avoid those areas where where the original problem arose, creating a phobic response in us.

A rich man stays and builds upon his good fortune (future hope). A poor man is always trying to escape his bad luck (past inheritance / negative attitude towards life).

Observations from another Planet

When your attention is directed or pulled elsewhere, than where you are, this is when accidents occur (other people's demands / events that grab your attention).

When something happens, we have two possible reactions - wonder what caused it and investigate or blame somebody else and stay in your comfort zone (it's their fault / responsibility, not yours).

Liars are always tense because they're always on the defense against external probing. Honest people are more relaxed because they've nothing to hide (no fear filled secrets / no embarrassing memories).

Pleasure isn't pleasure. It is the expectation that an experience in the future will be better than it actually is (time is the lure that draws us ever onwards).

The well are self-supporting. It is the sick in mind and body that need our help as well as their own, to pick themselves up, dust themselves down and start all over again, with their lives after an illness, accident or mental trauma.

The reason failure affects us so badly is because it stops us enjoying the pleasure of success (excludes not includes, leading us to withdraw in shame).

We treat others with disdain, to cover up our own sadness at past failures (put on a brave face). The more we fail, the less capacity we have for failure, so withdraw from further effort and treat with contempt the newbies, who in their innocence try, where we have given up.

Observations from another Planet

We all have our motives for action. It is when these clash that problems arise as we try to impose our sense of morality on those that have no problem with what they are doing, even if we do. Our perception and definition of what is good or acceptable, may just lie outside the scope of the other person's experience and so our reaction may seem inappropriate to them as theirs does to us.

What you have merged with, you don't notice as it has become a part of you. What you are separate from, you do notice (blissfully unaware versus brutally revealed).

Opinion is categorizing an experience as good or bad (wanted / not wanted), rather than accepting it for what it is as a physical effect.

If you love yourself, you'll accept everything else given to you with love. If you hate yourself, you'll feel too guilty (unworthy) to accept anything with gratitude and will be suspicious of such gifts.

Anybody with the patience to follow an event from beginning to end, will know when the mechanism that sets it off, has reached its climax. Anybody without that patience, will always be taken by surprise (not be ready for it to happen, having already abandoned it).

When we're introverted (depressed /ill) we cannot stand being rushed or moved as every sense is exaggerated. In this state we notice the smallest change but find the largest overwhelming. This is why we defend ourselves against intrusion from the outside world / insensitive others.

Observations from another Planet

While rhythm assists memory, it also puts us to sleep with its repetitive certainty, so that you fail to notice things outside your knowledge base. Melody wakes you up with its uncertainty because of its continual change of direction, leading you all over the place. The former is a habit (white-out) where you cannot distinguish one thing from another (all details lost in a general fog).

Accidents happen when we abandon control (relax / let go) as stress occurs because we hold on rigidly for dear life (fear change / uncertainty / chance).

People who want to be thought of as right, will correct others errors but they won't correct their own.

To solve a problem, you need time to explore possibilities (experiment). Willpower comes after because the focus required to act, makes you blind to everything but the task in hand (mentally disabled / physically enabled).

The closer you get to something, the finer your control over it grows. The further away you are, the less control you have (general awareness alone).

People with fixed ideas try to force things to work because they believe they should do so in that particular way (anger rush). People with open minds, stop and observe how things really work, without prejudice, learning through direct observation.

The more you resist letting go of something, the stronger its effects when you do finally release it.

Observations from another Planet

Followers will drink poison, thinking it harmless and vice-versa, simply because the government / religious leaders say so (self-hypnosis).

It is hope that wakes us up and despair that sends us back to sleep again.

To understand anything as a whole, you need to see both sides of the argument, not just one (no Yin without Yang).

Psychic power is not about what you do but what you don't do (what you censor about yourself - the limits you impose, that ensure you do not waste your energy on outgoings but save in the way of incomings).

The intelligent are explorers, whose journeys are self-led, not guided tours run by somebody else.

There are two types of people in the world - those that want to expose the truth and those that want to promote the lie (illusion).

It's not your position in society that makes you a saint or a sinner (criminal or law abiding) but your actions.

Fear of getting things wrong, leads us to get stuck in habit. Superstition and ritual are based on this fear of change (phobia against trying anything new / alternative ways of doing things).

Shallow people judge others by their mental prejudices, not by the physical evidence before them (reality, untested by them).

Observations from another Planet

The warlike are not interested in explanations, just action. The peaceful are only interested in explanations, not excuses for action.

The reason we have accidents is that our bodies are present but our mind (attention) is not.

Somebody else's life is like another culture, another country - to be ignored out of fear or explored out of curiosity.

Highly focused people don't see new possibilities, just old goals. This leaves their minds closed to change, not relaxed and open to it (receptive / perceptive).

We have an investment in making out others are wrong because it implies we are right (have the moral high ground).

Tyrants don't allow choice (originality) because they want obedience (conformity), instilled through fear, not autonomy (free will / individuality).

Failure makes you doubt your knowledge base. It's not that the former is anything to do with the latter but failure shatters your confidence, your certainty.

The emotions are a musical scale. They are not meant to be played as a one note samba.

Panic is failure to stop and look. It is fear driven escapism as opposed to calm assessment and decision making.

Observations from another Planet

Certainty comes, not from doubt (witness) or confidence (action) but experience (result of action or inaction).

Movement stirs things up - stillness (meditation / contemplation / thinking) allows things to settle down and become clear, instead of chaotic and confused.

Meditation is calming down and finding yourself inside your mind as action is speeding up and losing yourself outside, in the physical world.

The frightened demand obedience from others (physical movement). The calm demand answers (understanding from themselves / perception).

We get lost in action, when we want to avoid thinking about and feeling something that causes us pain and confusion, and yet facing it in silence and stillness is the only way to resolve it.

Lazy people tell stories, rather than put in the effort, to seek out the truth (find the cause behind an effect) as frightened people tell lies, to protect themselves from its expected consequences (other people's reactions).

All action leads to destruction. All thought to construction (slowing things down, forms reality (freezes it in time and space) as speeding things up, melts it into action (turns it back into formless energy).

Habit brings stability by removing doubt and confusion (continual change leads to chaos and disorder).

Observations from another Planet

A patient person settles down and waits because of his knowledge of the situation - an impatient one bursts into action because of his fears.

People with fixed ideas try to force things to work because they are sure their theory is correct. People who doubt and question, check their ideas against reality.

All movement is by its very nature, a violent breaking away from a settled (peaceful) state. It is the opposite direction from inward seeking of answers as it is outward implementing of them, through action (concentrated thought to dispersed energy - introversion to extroversion).

We are all responsible for our decisions, to act or not, when faced with certain situations. We cannot therefore blame others / the world in general, for what it does or doesn't do, only ourselves for our own actions and failures to act.

Awareness of detail is a sign of being grounded in the present moment as general awareness isn't. It is a sign of intelligence, in the sense of searching for answers and collating information, in order to make sense of the world.

Awareness of detail is a sign of peace in your life (self-control) as lack of it leads to chaos and confusion (conflict / violent eruptions of activity).

When you are charging into the future, following your imagination, how can you not but leave your memory behind?

Observations from another Planet

You cannot know the future, only create it through your actions. The past is based on previous experience (memory) - the future on imagination of what could be.

Those who want to solve problems, seek out the truth. Those who want to dump responsibility, do their best to bury it.

We blame things for not working in the present, when we forget the abuse we caused them in the past.

Responsibility is the silent pursuit of the truth. Blame is verbal 'pass the parcel' of it's not my fault this happened, it's someone else's.

When we are doing something for ourselves, we do a good job because we are conscious of the end result. When we're doing something because somebody else wants us to do it, neither our conscience or consciousness are engaged and quality suffers.

When we think that we are individual failures, we feel ashamed but finding it is a common fault that other people make too, removes this imaginary stigma from our lives.

We live for ourselves but die for other people.

Pain is stored, when ignored and drained when you put your attention upon it.

Prejudice isn't what you do - it is what you resist doing (experiencing).

Observations from another Planet

What we do is fact. Why we do it, is theory.

There are two ways of dealing with a problem - suppressing it or enhancing it. The latter drains the situation, the former allows it to build up (sweeping it under the carpet, versus exposing it and seeking solutions or at the least giving voice to fears and aspirations).

To make sense 'of' the world and to make sense 'to' the world, you have to control the urge to run 'from' the world (panic).

Instant reaction confuses. Who did that? Did I do that? This is because there is no discernible gap between thought and action. In the real world it's more obvious what follows what (cause and effect). I lay the fire - paper, kindling, coal. I strike a match and there you are.

Fear destroys bridges to the outside world. Intelligence is connecting the dots, to make sense of reality. Those we consider ignorant are shut off and shut out from this external awareness. To them reality is a confusing, chaotic mess on all sensory levels because they cannot distinguish individual differences. Those that can separate out life's components, can make perfect sense of the world and therefore can control it better than those who can't do this.

Awareness brings responsibility. Suppression of others and repression of the self is fear, stopping experience (learning, growing, discovering the new, the different, the strange).

Observations from another Planet

The purpose of education is to stop you making the same mistakes you made in the past, over and over again or duplicating previously executed work (reinventing the world) as an individual or society. Healing is helping you face and correct past errors, physically or psychologically. It's remembering or revisiting them and acknowledging (accepting) what happened, so you can put them right as far as possible, in mind or body.

The stronger your grip on reality, the slower your speed. The less hold you have, the quicker you need to react, to regain contact / control.

Protective violence is trying to create 'me time,' in order to sort out your life by pushing others away as being sociable is having nothing bothering your conscience or awakening your consciousness.

If you don't get enough sleep, you're not really grounded, which means you're not in control of your faculties (centred in in your body, inside looking out, rather than outside, looking in).

It's not experience that changes over time but our perception of it (overexposure drains it of effect upon you - underexposure keeps it alive in you).

True action is silent (does what it has to). Fear doesn't shut its mouth as it must interfere, which discloses its true identity.

Observations from another Planet

Fear is the brake and anger, the accelerator that drives our body the car.

You can either chase others to get their attention or give yourself respect and stop running away from the person who should matter most to you, which is you.

If you cannot understand how something works, it's magic to you. If you can understand it, it's science.

Anger and hatred shout. Love is silent because it is content.

There are two types of people in the world - those who want to understand how something works, so they can disarm it as a danger and those who want to destroy it, so it stops grabbing their attention as a problem.

The more you are aware of, the more you're response-able for, The less you're aware of, the less you're responsible for.

The frightened rush things because they are afraid that they don't know what they are doing. The calm do things at their own pace because they realize that at last, they do know what they're doing.

It's easy to give gifts - harder to give yourself.

Hate is the attempt to stop something - love, the effort to start it.

Observations from another Planet

Problems are always general - solutions are always specific.

Imperfection is the start of the journey - perfection its end.

PART TWO

(The politics of experience)*

*Ideas that are mostly set in short lived time

Bodies are what we are given as a canvas – belief is the paint brush we use to give ourselves direction in life (The body is place (constant) as the mind is time (change)).

Observations from another Planet

When we are following a trail, a storyline, we get lost in intense concentration upon it. The feeling of physically being lost or alone, comes when we move out of this state and view the everyday world again (dispersed attention / caught in the boring immensity of existence, where you end up feeling insignificant because you are (a real ego destroyer).

A slave state is thinking 'Thank God this isn't happening to me!' (A pecking order of hate, a hierarchy of fear). A free state is thinking 'Oh my God, that could be me!' (equality / humanity).

Intelligence's burden is hyper-sensitivity (great depth / awareness of detail). Lack of intelligence displays itself as shallowness (easily moved, like flotsam on a wave). To the former, the smallest disturbance is like an earthquake – to the latter, each earthquake is like a small disturbance (large hold on reality, versus little hold on it).

When you are in a witness state, you look on in fascination at what is happening, to see the end result. When you are in a participant state, you get angry and leap into action, in order to stop the process progressing any further).

When we start some task or journey, all we are aware of is its enormity, which terrifies us because of its gigantic scale in time and space. When we reach the end though, it is just a question of tidying up what's left (dotting the I's and crossing the T's) which is satisfying because it is small scale closure (the end in sight).

Observations from another Planet

Disappointment slows us down in our tracks, sabotaging our enthusiasm for life. In this state of political assassination, we are stopped totally in our tracks, mentally and emotionally murdered, which leaves us in a dark place, rather than burning with the light of own joy at realizing we are still alive and still have hope left of changing reality.

There are two types of people in the world – those who just want to rake over the coals of the past and those who want to start a new fire.

Arrival anywhere is is confusing and chaotic because it is overwhelming with its flood of sensations (input), be it a new job, home, relationship or life in general (heaven heals because it inspires as hell (the old) bores and leads us to abandon our lives – return to chaos and confusion as the former leads to a search for meaning (order and clarity) in our lives). This is why the end of something is more satisfying than the start.

It is the refusal to accept things as they are that makes us impatient and intolerant – leaping into action. It is a feeling that we don't have time to have our attention diverted from the task in hand, which we are rushing to complete (pressure / stress, from limitations of time, space, energy or matter (resources in the material world).

When we are in an active (doing) mode, we do not judge the results of our actions. Only when we are in a passive (thinking state) do we examine and take note of what we've done, to see if it's the desired end result we want or not.

Observations from another Planet

If you let things be, they're not frightening. If you try to stop them (fear their existence) then your concentration upon them, makes them fearful.

When we don't treat things with respect but just discard them after use, then we have no firm memory of where they went and therefore find it hard to recover them (order (control) creates memory – chaos creates confusion).

Replacing things where you found them, ensures memory through order, is retained. Using and then throwing things to one side, creates chaos for the body and confusion for the mind.

When we become disillusioned, what we saw once as magnificent or awe inspiring, suddenly seems dull, mundane, obvious and our interest in it dies – be it a job, relationship, home or life altogether. When this happens, we drift away from this situation, seeking a new mystery to explore, a new way to live.

Critical people want others to conform in appearance or behaviour to their yardstick of physical or mental perfection. Uncritical people accept others for what they are, warts and all.

The reason habitual abusers are not caught more easily is for the same reason castles have torture chambers – the crime is hidden behind a facade of quiet normality.

The traumatized compartmentalize their experiences, locking them up in prison cells, to avoid experiencing and integrating them. They therefore are concentrated emotionally, unlike free memory, which is more dissipated energetically.

Things only affect you, if you move physically or are moved emotionally.

Force moves things but stability comes from presence (the build up of resources instead of their dissipation through use – mental, physical or emotionally.

You don't forgive others sins for their sakes but your own, just as you can only forgive your own mistakes and regret them, when you see how they impact upon yourself in a negative way. We also don't need God to forgive us or society to punish us for our crimes. What we need is to have our conscience awoken or to gain conscious awareness of how reality really works.

Opportunities rush passed us like driftwood in a river. They exist in time which flows by us in a constant stream of ideas and images. If we do not grasp them when the chance arises, they slip into the past and are lost forever.

If we see things as a gift, we accept them. If we see them as a curse, a threat, a distraction – we try to suppress their effect upon us, rather than relax and enjoy the experience.

Fact is what happened – theory is the attempt to explain the past or predict the future.

Observations from another Planet

The Micawber Equation says we are all responsible for our own existence. If we expend more than we take in, we drain ourselves. If our input is greater than our output, we become bloated with unused resources. If these two balance out, then our life itself is balanced out.

When pessimism speaks, it lies (falls short of reality). When optimism speaks, it lies (overshoots). Only intellect (measured, neutral response) is accurate.

The paranoid think everything is deliberate – the reasonable know it isn't.

Those who proclaim how hard done by they are, aren't. Those who suffer in silence, are (look, don't listen and believe).

Disability – it's not what's wrong with your body but with your mind, that holds you back.

The insane in their arrogance, think they're sane. The sane in their innocence think they're crazy.

Love settles and builds – hate destroys and moves on.

Anger is feeling you have no time to think or space to act.

What drives extinction is rarity (the less there is of something, the higher the price paid for it).

Observations from another Planet

When something happens, ask yourself – is this new life trying to get out or old death trying to get in?

We've become a world based on meaningless qualifications, rather than meaningful qualities.

If you make a man angry, you arm him with hatred. If you make him laugh, you disarm him.

People who rush headlong into the future, leave themselves and their senses behind.

Marriage is a social / legal convention. Living together is a choice.

There are two types of people in the world – those who assess the evidence and adjust their beliefs accordingly and those don't let facts get in the way of their prejudices.

Skilled workers are never cheap and cheap workers are never skilled. What kind of society you have will depend upon which of these options you chose – technically advanced ones, run by free men or slave states, run by slave labour.

Scared people don't think and thinking people don't scare. Calm people stop and look – panickers run away without a backwards glance.

If you want an easy life, follow the materialist path. If you want an easy conscience, follow the spiritual life.

Observations from another Planet

Anything that blows up the ego, can blow up the world.

Death is what separates us – life is what unites us.

We are either in debt (have financial diarrhoea) or hold onto and accumulate funds (become financially constipated or obese). Balance is the point in-between (neither rich nor poor (input equal to output).

In peace everybody matters – in war nobody does.

To see is to be lost in eternity (bored / stuck). Music (sound) breaks up time into into rhythm and melody – that is predictable repetition / a road to follow.

Boredom eating can be overcome by replacing visual entertainment with music because sound leads to movement as sight does to stillness.

Materialism is an addiction to things and the rich are its pimps and pushers. They too are addicts to wealth, power and position – fighting to keep their addictions supplied, to the ultimate level and quality.

Money has no conscience to spur it into action. It cannot pick up a shovel and do manual work. Yet people think it can cure all ills, so pour loads of this inert matter at a problem, thinking and hoping it can inspire action to solve the situation. They forget spirit and intention, instead relying on a subjective belief in finance and greed, to buy their way out of difficulty, through trade-offs with others.

Observations from another Planet

Resistance, that is rebellion, fear, rejection, separation ('No!' attitude) wakes us up as non-resistance (acceptance, unity, courage, conformity ('Yes!' attitude)) puts us to sleep: Settling versus leaving.

A university degree doesn't make you a scientist, anymore than going to to art school, makes you an artist. It is the urge to create or discover that makes you either of these.

If you're not expecting to be attacked, what good is a gun in defense? You can be blown up, stabbed, bludgeoned, shot before you can react. Did armed security stop Kennedy's assassination, the attack on Ronald Reagan or Lee Oswald being shot? Guns are only good as offensive weapons, not defensive ones, unless the attacker is a poor shot or has multiple victims to aim at, in which case counter offensive measures come into operation. In truth the best defense is vigilance, not weaponry.

As a sense, sight is about stopping and looking as sound is about listening and moving.

To hunt for food is necessary for survival. To hunt for trophies with a rifle is cowardly as you're killing from a distance, an animal that cannot defend itself and may not even know that you're there because of this gap. Is it really anything to be proud of (assassination without risk)?

Conscience is caring about when you die and what you leave undone. It is not about the gardening, housework, the work you did for somebody else but whether you are proud of the effort you put into your life and whether it will leave a

106

long lasting impression / legacy or be instantly forgotten as trite, mundane, poorly done.

There are two types of people in the world – those who think they're like everybody else and those who think they're nothing like anybody else.

Knowing somebody is worse of than you, makes it easier to complete difficult tasks (makes you a missionary). Knowing somebody is better off than you, makes you a victim, full of resentment (rebellious, not obedient).

We travel in hope but arrive in despair, when we realize it is the same place we left earlier (the circular nature of time / the boredom that comes with exploration and the reduction of the unknown to the known).

If you feel, you don't think. If you think, you don't feel.

If you question whether you are something or not, you aren't at that point. If you defend yourself against something, it is because you know that you are it or fear that you are and don't want to be associated with it (in denial). For instance if you think you are stupid, you're not because you have the right attitude of doubt, which is the willingness to learn (change). Likewise if you think you are smart, you are not because you have stopped questioning the validity of your viewpoint (become arrogant / certain) – no longer wishing to explore the different but instead sticking with the known, the same.

Government (the law) is about quality control.

Observations from another Planet

Sadness is acceptance of defeat as anger is rejection of it. Both take it as a serious loss. Humour is laughing off failure as unimportant and getting on with your life.

Putting things in order helps you discover what you've got. Throwing things into a disorderly heap, leads to them being buried in a confused mess and items remaining unseen beneath others.

In childhood everything overwhelms us through underexposure as in old age it underwhelms us through overexposure.

Deprivation sensitizes us to the sensory world as overexposure desensitizes us to reality (white out effect at the poles as opposed to desert oasis).

Ignorance frees us from interaction as resistance ensures it.

Every blow to our ego is a blow to the rhythm of our life. It literally knocks out the light in our lives too, which is why depression, anger and not knowing are associate with blackness. Light is generated by forward momentum as with a dynamo and the dark created by stillness (stopping everything – the silence of death).

Intellectual pursuit requires deep concentration (stilling body / silencing mind, in order to see the truth of what was (past memory)). Emotional pursuits require dispersed attention – that is active body, reacting to external (perceived) reality.

Observations from another Planet

When we feel, we can do no wrong (are a success), it inflates our ego (fills us with pride). When we feel, we can do no right (a failure) it deflates our ego (shames us).

The reason we don't feel major injuries but moan about minor ones, is that we are conscious of the niggling little things but are knocked totally out by the larger ones (unconscious, so we cannot sense or react to the damage we suffer or if we do, it is primitive convulsions and sounds that come out, not more controlled responses.

Those who lead full, busy lives, have no time to learn anything new (no room to change). Only the relaxed (bored) can fit in the new and motivated to do just that.

In an emergency, everything gets honed down into the simplest form possible as this increases the speed of action. In peace and calm, everything develops into more complex forms as there is time to explore all possibilities. This explains the contrast between the two states.

If half the world stopped interfering in the other half's life, the whole world would be a better functioning place.

The trouble with the insane is that they don't know that they are insane, so cannot see or understand the limits of their condition (no insight).

Cowards need to justify their actions (claim to be victims) by blaming others. The courageous just get on with their lives, by proudly being themselves (an example to all).

Observations from another Planet

They include everyone in their world, without the fear and prejudice that drives the insane.

Religion that preaches greed and hate is not religion but politics (physical revolution, not spiritual evolution – division of bodies, not unity of souls: True religion doesn't limit or exclude – that's politics.

Women create the race – men create short lived dynasties.

There are two types of people in the world – those who obey the rules without question and those who question the rules without obeying.

Action, by its very nature, is destructive – either accidentally or deliberately. Inaction creates the peace for things to grow in (room for ideas to develop / the space to act in).

Faith builds mountains – despair pulls them down again.

We're not willing to learn anything new, until we get fed up with the old.

Life is magic – death is disillusionment with the show.

There's no point obeying the rules, if the rules don't work.

We judge something as obscene – not because it doesn't exist but because we wish it didn't.

Observations from another Planet

When adults act like children, then children are forced to act like adults.

Breaking up things into smaller units (not too much or too little – the optimum amount in other words) makes them easier to deal with. Eternity and infinity overwhelm us with their immensity, unless we reduce them into manageable chunks (measurable distances, through time or across space).

Impatience and intolerance as tactics are the attempt to cut things back to basics / remove distractions (simplifying / shrinking reality).

The deaf appear stupid to those that can hear because they don't have the feedback capabilities, to sharpen the delivery of sounds, in other words they cannot pronounce words clearly and distinctly. The assumption is that sharp reproduction, shows tight control and that in turn shows skill (intelligence, through repetition of actions): Sign language would disavow this mistake because speed would increase in the chosen medium of visual communication, where no handicap exists.

Open prisons should be reserved for white collar and petty criminals because if they escape, they are no danger to society or less so than violent criminals, who are habituated to aggressive actions to achieve their aims.

Sex is surfing on the sea of life.

Observations from another Planet

Medicine is not for making money but as a service to the community (getting sick and injured workers fit enough to be able to serve society again).

Intelligence is the ability to differentiate between one thing and another. Lack of intelligence manifests in the ability to make distinctions.

High morale goes with high intelligence because it stimulates the urge to find out things as low morale goes with the opposite (no curiosity, no urge to explore - apathy, depression or outright fear of going beyond your limits, that is feeling, being aware, being response-able).

Death for the mind or spirit, is when you cannot imagine a future, don't want to remember the past and don't want to perceive the present.

There is a sliding scale between life and death, that includes motion for the body, emotion (energy) for the spirit and knowledge for the mind. The more alive you are, the more stimulated into thought and action you will be, plus the more energy you will generate, dynamo style.

If you cannot face the truth about yourself, how are you going to recognize it in others? The growth of intelligence comes from the subsidence of fear, when you stop and observe rather than run and hide (avoid contact with the outside world).

People who don't use their senses to check what they are doing, blame the outside world for their failures (too quick to act - not slow enough to observe the truth and learn from it (shallow, not deep thought)).

The trouble with ingenious solutions is that they sometimes have to be used by people who are so physically tired that they can only deal with simple solutions.

If we truly love and admire someone or something, we stand back and let it develop. If we fear where it is going, we step forward to stop it or change its direction.

Stress is trying to do too much with too little - be it time or resources.

Determined people can make something from nothing (are creative). Negative people can make nothing from something (are destructive in mind, body and spirit).

Positive people can turn anything to their advantage (increase potential in everything they find). Negative people reduce the potential for growth in everything they come across, by thought or deed (destroy confidence / smash opposition / reduce reusable material to rubble).

How can you perceive anything in depth, if you don't settle and observe? Movement destroys clarity and removes certainty therefore.

Proximity leads to clarity because the closer you are to something, the more detail you can see and the more certain you are of what you are seeing.

Observations from another Planet

There are two search modes we all operate in - internal (memory) and external (perception).

Immature people don't try to find solutions to make their lives better but instead prefer to moan about how imperfect things are for their needs.

Victims only see problems – victors only see solutions.

If you think about those that are better off than you, you will always be jealous and miserable. If you think of those worse off than you, you will always be grateful for what you have.

Courage is trusting in the good (connecting) as cowardice is distrust (disconnecting from the world / withdrawing from it).

Cowardice is fear of taking risks.

Terrorists are not afraid to die because death is certain (predictable). They are however afraid to face life because it is uncertain, unpredictable.

Idealists are willing to die for their beliefs – realists are willing to live for theirs.

Resistance prolongs an act and therefore the agony because it cannot be accomplished quickly and smoothly, in the way that compliance can (accepting your fate, when there is no alternative available – that is overwhelming forces at work in a chain of events).

When we don't have the time and space to develop our individuality, we become emotionally and intellectually crippled (deprived).

When we think life is unfair, we slow down our efforts and decrease our output (self-sabotage). When we think it is fair, we increase our output, act enthusiastically and make a good job of what we are doing (develop skill in our efforts).

Memory, like perception, requires stillness and silence to create clarity of thought and vision.

The past is certain, so is easy to define and discuss. The future is fluid, so cannot be defined and discussed (yet to be set in stone).

Evolution is before and after movement, which changes state from one set form to another, like a magic trick. Looked at statically you only have before and after as certain. Between is blurred by activity (transition / decision / choice).

People who lack confidence will feel offense at remarks because they go into defense mode. This is a sign of low self-esteem and guilt (feeling an imposter / out of place). Those with high self-esteem laugh off all remarks as harmless because they are aware that these are reflections of somebody else's negativity and have nothing to do with their own sense of self (outgoing positivism, not introverting fear of discovery at their own unworthiness).

Observations from another Planet

Apathy allows us to withdraw from life by accepting defeat. It lets us regroup, gather new resources, think out new strategies and join the fight for life again at a later date, when we are more ready to deal with it. It is the sleep of death that allows rebirth another day.

Life is a wave of energy passing through the universe. It moves from past to future, revealing a new world to us, until like an empty rocket we fall back to where we came from, exhausted of energy, like an empty vessel.

When you are aware of the danger you are in, you freeze. When you're unconscious, you act without a second's thought, even if it kills you because you don't see the situation you find yourself in as real.

We lie to avoid the consequences of our actions and tell the truth, to salve our conscience: The first is done in fear of other people and the second for our own sakes.

Everything has a beginning, middle and end. It is always hard to start something because of its immense size (the whole incomplete thing / the untouched future). It is easier to finish because you know you are approaching the end (it's shrinking to nothing stage) and the half way point is looked at and worked towards because you know after that juncture it is always downhill (less in time and matter, to be dealt with).

Ignorance is a choice, not an inability or disability. It is the decision to pretend something doesn't exist by shutting out awareness of its existence.

Observations from another Planet

When we become self aware (gain a conscience) that is when we restrict our actions (become ethical as an individual or society).

People who are nervous or rush around excitedly, only see the parts of reality that they concentrate on momentarily. Those who relax, step back and take in the larger picture (the whole).

Speed disturbs through disruption as stillness calms through stability (constancy).

Noise destroys concentration as silence creates it (allows us to settle).

Death is abandonment. It is letting go of the heavy hold on life we have and floating free instead. To live is to put effort into life – fighting to survive. Death is saying it's not worth the effort and giving up on a job, a marriage, a town, a life and moving on.

Paranoia is simply being separate from the world and seeing it as hostile for this reason (unpredictable, unknown, strange, different from the self).

Respect is awareness. It is waking up to your own stupidity (self-destructive behaviour, in removing from your life, the survival mechanisms you need to continue in this world, this life).

We consider something a mistake, when the outcome is not what we wanted or expected but it can lead to the discovery of something new, that we did not know about the thing we were dealing with (a different facet). This could be of use to us at a later date or at least of interest now.

On the most basic level masculinity is denial of femininity.

Crime is refusal to accept responsibility for your actions or those of others around you (blame culture).

Nobility is the acceptance of your fate, no matter how unjust. Immaturity is rebelling against a situation you find yourself in, even if just (in other words you are actually guilty of the accusation against you).

When under pressure we are more likely to become frustrated and violent, rushing things. Peace is slow and deep (no pressure / total control).

Accidental discoveries are made when deliberate actions are forgotten about.

False association syndrome is where you think something is the same as something else because it is similar in form or appearance. It is based on vague memory and vague perception. It is not the same as false memory, which is based on suggestion and imagination (self-hypnosis replacing the truth of an event, totally). This is partially true in detail but mistaken in whole identity - facts right, conclusion wrong.

Insanity is the unconscious self-sabotage, of the future of the human race, through individual suicide, social group rebellion and warfare of nation against nation.

Observations from another Planet

The difference between sport and art is that the former is competitive (one individual or team versus another) and the latter cooperative (a group working towards the same end or an individual trying to perfect their skill).

You can tell an addict because they can't bare to be separated from their addiction (get violent when they are frustrated at not being able to obtain instant gratification for their addiction).

Openness and sharing makes something less addictive as rarity and protectiveness, makes it more so.

Speed is a sign of addiction because it is based on fear of not experiencing something or losing it altogether.

Self-importance and indeed a belief in importance at all, is a sign of fear of loss (addiction) as seeing everything as unimportant, is a lack of it (apathy - withdrawal of contact from the addictive source).

Pessimists over prepare for possible dangers - optimists don't prepare at all, so have to improvise or die from lack of preparation.

Arrogance doesn't learn because it thinks it already knows the answer. Humility discovers because in its innocence, it explores in order to find what truth is / what works.

The belief in punishment comes from the idea that something was done deliberately against you (ego / negative feedback - that is victim mentality). Forgiveness is the realization that it's all accidental and nothing to do with you at all (nothing personal - victor mentality over the self / positivism / release from slavery of the mind (ideas / beliefs)).

Observations from another Planet

When you don't want to do something about a situation (laziness / cowardice), you will ignore all the evidence that it needs to be done.

Every time you open up to something new, you learn about it in more detail and gain more control over it. Every time you shut out awareness of something and refuse to deal with it, the more you leave it in stasis as unfinished business and the more power it exerts over you.

To spoil a child, give it all it demands materially. To make a child, give it all it needs spiritually.

Could it be that the reason the Near Death Experience appears so wonderful, is that like an LSD trip, it is a return to the pristine state of childhood - free from the psychic armour that protects us from the negative effects of experience (loss).

If you want people to follow you into the future, ignore them. If you want them to stay with you in the present, pay attention to them: the first creates dependency (addiction), the second a relationship (mutual honesty through exchange and exploration of differences).

Apathy is inaccurate and accident prone because it doesn't care. It relaxes out of a situation as anger and conscience crashes or squeezes back in, to control or take back control of a situation. It is conscience (conscious awareness of the consequences upon the self) that takes pains to get things right I.e. complete actions / be accurate. It suffers for its art as apathy avoids suffering by giving up the pain of trying (death of his spirit).

Negative people make life a burden for themselves and others. Positive peace remove all pressure from themselves and others.

Perception is the superior faculty because it is in present and of the present. Memory is inferior because reality is subject to change and therefore can become redundant as it is of the past. Imagination too lacks substance as it is what is hoped for, not what has come to pass.

The longer you are somewhere, the more you revert back to old habits (introvert) as being lost in the thrill of the new, extroverts our attention instead (imagination wins over memory / hope over despair).

The new extroverts our attention - the old introverts it. In the former we get dragged out into the macrocosm (general awareness). In the latter we get sucked down into the microcosm (specific details of existence).

Memory is a short cut. Without it we go into search mode, seeking answers instead of using ones we have already and can work quickly and fluently with.

Optimists adapt to change (loss) - pessimists are crushed by it.

Optimists are excited by the idea of change as pessimists are frightened by it.

Pessimists start off with an answer - nothing can be done. Optimists start off with a question - what can be done?

Things get lost in a sea of action and found in a pool of stillness.

Age sees because it slows down - youth doesn't because it speeds up (ignores).

Only the present is accurate - the future exaggerates and the past underestimates.

What if the world is held together by willpower and molded into shape by strength of mind?

What we are afraid of, we try to get away from, if external and try to oust, if internal.

Prejudice is not really about race, religion, sex or sexual persuasion but exploiting a weakness, which is feeling ashamed of who we are, what we do or what we are associated with. Those who are proud of themselves, cannot be emotionally manipulated or blackmailed into submission as pride needs no defense or excuses for its action. Its attention is turned outwards into the world, not inwards towards the self through fear (self doubt / hypocrisy / lack of authenticity).

Order and clarity are a sign of somebody who is high functioning (positive, elated, creative, heightened emotions). Chaos and confusion are signs of somebody, who is low functioning (negative attitude towards life - apathetic, depressed, distrustful, rebellious).

The outsider view of reality is inferior because it is based on theory, not fact. Only the insider view is superior because it is based on direct knowledge as opposed to guesswork (cause not effect, memory (personal history) not imagination).

Fear leads to strong time management and planning for the future - hence the military trying to frighten and bully the life out of recruits. A relaxed or apathetic state, means either having all the time in the world, in which case you laugh at fear and limit or back out of life and do nothing, which makes action and planning nonexistent - that is chaotic and confusing because you are not prepared for change / disaster.

Memory and routine are synonymous because repetition breeds certainty, just as a large emotional impact ensures long term memory as well.

To exercise (move) the body and ignore the mind (inner world). To exercise (move) the mind, ignore the body (outer world).

Memory is dependent on interest or impact, forgetfulness upon disinterest and lack of impact.

Observations from another Planet

Idealists plan for an outcome they want. Realists plan for how things are; the former is let down by false expectations as the later is relieved when things work out positively.

To stop and face the past and confront the future, is a sobering moment, where you assess the direction your life has taken and where you want to take it next. When you are not doing this, you are in limbo, observing / learning but not putting that knowledge to use (uncommitted / hiding in the background / avoiding action - staying in the land of theory, rather than turning ideas into concrete facts).

Alice down the Rabbit Hole effect, is where everything becomes jumbled up in your mind because you are in a state of dispersed attention. When you're in a state of concentrated attention, everything in your mind is ordered, sequential and accurate as humanly possible because you are grounded in the here and now, not lost in the there and then (past regret / future hope).

Prejudice never learns anything new because it is too busy protecting the old ways of doing things / viewpoints. This is because the ego is heavily invested in being right and everybody else being wrong.

We are never aware of why we start something because it is a spontaneous reaction but we are aware of why we want to stop it as this is a conscious decision (slowing down and examining our conscience and consciousness as opposed to speeding up and acting without thought of the consequences upon yourself or others).

Why is it that those who have the least to say, insist on shouting it the loudest?

As we grow older, we gain more control over our lives and it becomes more orderly, more organized – replacing the spontaneity of youth (inexperience / lack of knowledge) with routine, habit.

The ignorant are easy to manipulate (be lied to) because they don't know what the truth is, so readily accept as the truth what may not be so.

Delicate (fine) control comes from being there (present, no matter what you are facing). Rough control comes from fear (tentative grasp on reality because you are reluctant to deal with what lays before you): Male energy / attitude as opposed to female.

The value of something is measured in your desire to have it or your fear of losing it.

Fear is the darkness that absorbs your soul – light, the courage that radiates from your spirit.

LIFE-LAWS

The human world works on equally understandable laws to the physical universe. To obey those rules leads onto success – to disobey them is to sabotage your own future and that of the world you live in.

Time is as important to the mind and human relationships as it is to science and the physical world. Short (fast) time is destructive to physical reality (thoughtless action) as long (slow) time is constructive (thoughtful action – time in which to observe, think and build in depth, through lessons learned by experience). Patience thinks nothing is important, so gives up and tries again when it fails. Impatience thinks everything is important, so refuses to accept failure, not backing out of a situation but continues to bash its head against a brick wall, in a self-destructive act, ignoring reality.

The truth is that we need to step back and observe the effects of our actions, to learn from them and forward to act and change the world around us, not ignore it to our cost.

When you realize the impact you've had on other life forms and reality itself, that is when you develop a conscience and want to leave the world. As long as you think of yourself as innocent, you want to stay

AFTERWORD

This book and the preceding one were written for the benefit of you who will come after us on this planet but in all probability-

"There is nothing I could have told you
that you'd have understood
And there is nothing I could have told you
that would have done you any good"

The ballad of Ghengis Smith
by Roy Harper

In case that is not true, check out the following websites:-

http://deepthought2.weebly.com
http://shallowhumour.weebly.com
http://flyingsaucery.weebly.com
http://logiclistsenglish.simplesite.com

CPSIA information can be obtained
at www.ICGtesting.com
Printed in the USA
LVOW13s2017260617
539424LV00017B/371/P